the new brain

01000101011100101010000100101010000010101101001010000100111

the new brain

HOW THE MODERN AGE
IS REWIRING YOUR MIND

RICHARD RESTAK, M.D.

BEST-SELLING AUTHOR OF *MOZART'S BRAIN*
AND THE *FIGHTER PILOT*

RODALE

© 2003 by Richard M. Restak, M.D.
Cover Photograph © by David Roth/Getty

First published 2003
First published in paperback 2004

Printed in the United States of America
Rodale Inc. makes every effort to use acid-free ∞, recycled paper ♻.

The illustrations on pages 1 and 178 are © by Molly Borman.

The photo of Paul Bach-y-Rita's Tongue Display Unit on page 161 is used courtesy of Bill Angelos.

The figure on page 203, by J. A. Bates and A. K. Malhotra, is used with the permission of the journal *CNS Spectrums* and MBL Communications, Inc.

Book design by Christopher Rhoads

Library of Congress Cataloging-in-Publication Data

Restak, Richard M., date.
  The new brain : how the modern age is rewiring your mind / Richard Restak.
        p.   cm.
  Includes bibliographical references and index.
  ISBN 1–57954–501–7 hardcover
  ISBN 1–59486–054–8 paperback
  1. Neurosciences—Popular works.   2. Brain—Popular works.   I. Title.
QP355.2.R47   2003
612.8'2—dc21   2003005773

Distributed to the trade by Holtzbrinck Publishers

2   4   6   8   10   9   7   5   3   1   hardcover
2   4   6   8   10   9   7   5   3   1   paperback

To my wife, Carolyn

Acknowledgments

Thanks are due to the following for discussions, suggestions, and helpful criticism: Paul Bach-y-Rita, Thomas H. Carr, K. Anders Ericsson, Jordan Grafman, Joshua D. Greene, Pat Kuhl, Daniel Langleben, Jonathan Mann, Marina Nakic, V. S. Ramachandran, Mriganka Sur, Daniel Tranel, and Ronald Van Heertum. With special thanks to my editor, Troy Juliar, and agent, Sterling Lord.

"Does use and exertion of mental power gradually change the material structure of the brain, just as we see, for example, that much used muscles become stronger? It is not improbable, although the scalpel cannot easily demonstrate this."

—Samuel Thomas Soemmering, 1791

101011101011001010000100111**X**00000101011010010100001001

Contents

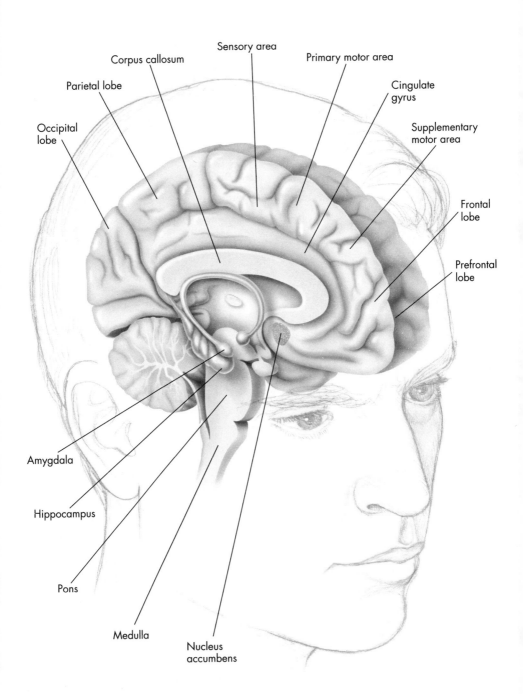

Occipital
lobe

Parietal lobe

Corpus callosum

Sensory area

Primary motor area

Cingulate
gyrus

Supplementary
motor area

Frontal
lobe

Prefrontal
lobe

Amygdala

Hippocampus

Pons

Medulla

Nucleus
accumbens

Introduction

We have learned so much about the human brain during the past two decades that it's fair to speak of a revolutionary change in our understanding. The era of the *Old Brain* is giving way to that of the *New Brain*.

The Old Brain was remote and mysterious, deeply hidden within the skull and inaccessible except to specialists daring enough to pierce its three protective layers. Thanks to that inaccessibility and the risks involved in plumbing its depths, brain experts knew little about the functioning of the normal brain; they certainly searched in vain for answers to such fascinating questions as, "How is the brain related to our everyday thoughts, emotions, and behavior?"

The New Brain, in contrast, doesn't require dangerous intrusions but can now be depicted using sophisticated computer-driven imaging techniques with abbreviated names like CAT, PET, MRI, and MRA. These techniques reveal exquisitely subtle operational details and provide windows through which neuroscientists (brain scientists) can view different aspects of brain functioning without opening the skull or performing other risky procedures.

Thanks to the development of new imaging technologies, brain science is capable of providing us with insights into the human mind that only a few decades ago would have been considered the stuff of science fiction. We can now study the brain in "real time" when we're thinking, taking an intelligence test, practicing a craft, experiencing an emotion, or making a decision. Brain tests can even indicate when we're telling the truth, as well as provide a quick estimate of our intelligence and specific abilities.

Neuroscientists refer to this new field as cognitive science: the study of the brain mechanisms responsible for our thoughts, moods, decisions, and actions. Cognition has been defined as "the ability of the brain and nervous system to attend, identify, and act on complex stimuli." More informally, cognition refers to everything taking place in our brains that helps us to know the world. Included here are such mental activities as alertness, concentration, memory, reasoning, creativity, and emotional experience.

In the era of the New Brain, the emphasis is shifting from diseases and dysfunctions to an understanding of the brains of the average man and woman. An exciting consequence follows from this new emphasis on the normal brain: Research can provide us with useful guidelines about our everyday lives. For instance, recent findings (discussed in chapter 1) indicate that by following certain brain-based guidelines anyone can achieve expert performance in sports, athletics, or academic pursuits. Such findings, of course, run counter to the traditional theory that sports achievers and geniuses are born not made, that our genes and other factors outside of our control impose limits on our individual capabilities. Not so. Instead, it's now clear that by learning about and applying this new research, most of us can reasonably expect greatly enhanced personal levels of achievement.

As another example, we now have good reason to believe,

based on brain research, that harmful effects on our brain can result from frequent exposure to graphic scenes of violence. Moreover, it doesn't seem to matter if the violence is fictionalized, "real life," or a combination of both (i.e., docudramas featuring depictions of violence based on actual events). Watching media violence changes our brain in harmful ways that we are only recently beginning to understand.

While this is not intended as a "self-help" book, I believe a lot of contemporary brain research has practical applications that can be put to use in our daily lives. Throughout the book I will discuss this research in sufficient detail that you will be in a position to decide for yourself what, if any, practical applications ensue in your own life. Included here are such fascinating areas of cognition as:

- Understanding the effects of media and technology on our thoughts and emotions
- Estimating the effects of stress on brain function, with an emphasis on the use of sophisticated instruments that can help predict those people who are at greatest risk for harm
- Formulating new brain-based ways of thinking about variations from normal behavior, such as Attention Deficit Hyperactivity Disorder (ADHD) and Obsessive Compulsive Disorder (OCD)
- Developing methods for enhancing our sensory capacities by harnessing the brain mechanisms involved in translating information from one sensory channel into another, such as the transformation of touch sensations into forms of visual perception

Twenty-first century discoveries about the brain will provide us with new insights into our behavior, thinking, and feelings. Thanks to technological advances, neuroscientists are already successfully correlating brain function with personality; synthesizing "designer drugs" for individualized treatments of patients suffering from depression, anxiety, and other neuropsychiatric illnesses; and correlating defective genotypes with violent or antisocial behavior. Thanks to such advances and the promise of even greater ones in the near future, it seems fair to say that technology, rather than biology, will play the major role in the evolution of the human brain. Over the course of this book, my goal is to provide you with an overview of the kinds of changes that you can expect to come about in the era of the New Brain.

chapter 1

Brain Plasticity: Your Brain Changes Every Day

One pivotal concept underlies our understanding of the new brain: plasticity. Basically, plasticity refers to the brain's capacity for change. As recently as only a few years ago, most neuroscientists believed that brain plasticity largely ceased by adolescence or by early adulthood at the latest. At this point the brain became fixed in its structure and function—at least that was the prevailing assumption. But this assumption, too, has turned out to be wrong.

We now recognize that our brain isn't limited by considerations that are applicable to machines. Thoughts, feelings, and actions, rather than mechanical laws, determine the health of our brain. Furthermore, we now know that the brain never loses the power to transform itself on the basis of experience, and this transformation can occur over very short intervals. For instance, your brain is different today than it was yesterday. This difference results from the effect on your brain of yesterday's and today's experiences, as well as the thoughts and feelings you've entertained over the past 24 hours. Think, therefore, of the human brain as a lifetime work in progress that retains plasticity—the capacity for change—as long as the "owner" is still alive.

While plasticity is easy to observe in the infant and child (an increasingly larger brain marked by an evolving complexity in the patterning of its cerebral hemispheres, accompanied by an increasingly sophisticated behavioral repertoire), plasticity in adults requires more subtle observation. The adult brain of a 20-year-old may not appear significantly different on examination from the brain of someone three decades older. An appreciation of plasticity, therefore, depends on the skillful use of neuroimaging.

The Basics of Imaging Technology

Overall, it's useful to divide imaging technologies into two classes: scans that provide the anatomic details—e.g., the brain's structure—and others that provide information about function—i.e., what is actually happening within the brain. Think of it this way: You are reading this book in a room that can be described in "anatomic" dimensions (height, width, etc.), but a complete description would also include functional elements: the ambient light, the background noise, the temperature of the room, nearby distractions, etc. Brain imaging techniques are like that too: either primarily anatomical or functional.

Thanks to imaging techniques, we can now explore what is actually occurring in the brain as we go about our daily lives. Take, for instance, the learning of a new skill. Let's confine ourselves, for the moment, to something relatively straightforward, such as learning a simple sequence of finger movements. At first the movements are likely to come slowly, but with practice, performance improves until the movements can be done with a minimum of effort. What is taking place in the brain during the learning of such an exercise?

To find out, Leslie G. Ungerleider, chief of the laboratory of brain and cognition at the National Institutes of Health, carried out weekly functional magnetic resonance imaging (fMRI) on volunteers while they learned a sequence of finger movements. Within three to four weeks of training, she could discern changes in activity patterns in three progressive parts of the brain: the prefrontal cortex, which is responsible for the intention to carry out the movements; the supplementary motor cortex, responsible for organizing the sequencing and coordination of the muscles involved carrying out the action; and the primary motor cortex, the part of the brain that gives the "orders" for the movements to take place. This progression, incidentally, conforms to the general principle of operation within the brain: the establishment of action programs (via the prefrontal and supplementary motor cortex) followed by the actual carrying out of the action (via the primary motor cortex).

Ungerleider speculates that during practice periods, new neurons are recruited into the neuronal network responsible for the movement sequence. At first only a few neurons are involved in the process, with more coming on board with increased practice. Most interesting of all, these brain changes could be detected a year later, even in the absence of further training.

If You've Seen One Monkey You Haven't Seen Them All

In a moment we'll take a closer look at the formation of motor programs for more demanding tasks like athletic or musical performance; but for now let's stick to more mundane activities.

For instance, remember the difficulties you experienced in high school while trying to learn a second language? Now those difficulties can be explained neurologically: the sounds (phonemes) of that second language came into direct competition with the phonemes of your native language, which were encoded in your brain's circuits over many years. In order to become fluent in that second language, your brain had to establish totally new circuits.

And although new circuit formation can occur at any time in your life, it's harder as you get older. In fact, the brain can most efficiently and easily encode phonemes from different languages during infancy. Indeed, we are born with the capacity for processing any language in the world! Scientists know this because of some clever infant experiments.

In one such experiment, a neuroscientist outfits an infant with a cap containing 20 electrodes that record the brain's electrical activity. With the cap in place the infant listens to the sounds of different languages played in the background while the researcher monitors its brain wave patterns. As the recording shifts from one language to another, the electrodes detect changes in electrical patterns in the infant's brain. These changes occur even for languages the parents do not speak and that the infant has never heard before. Gradually the infant will lose this remarkable ability to detect changes in all languages except the language he or she hears on a daily basis.

Researcher Pat Kuhl, in the department of speech and hearing sciences at the University of Washington in Seattle, has shown, for example, that Japanese infants can easily discriminate between American English *r* and *l* sounds, as in the words *rook* and *look*. Adult Japanese, in contrast, have great difficulty making *r* and *l* distinctions (there is no *l* sound in Japanese). After its first year of life, the infant loses this ability, as measured by the brain wave patterns, which no longer vary when an *l* sound is followed by an *r*, or vice versa. That's because the infant begins to process only the language spoken by its parents. In the absence of continued exposure to another language, infants lose the ability to detect crucial sound differences between their native language and other languages.

But adult Japanese can reacquire this ability, according to James McClelland, a professor of computer science and psychology at Carnegie Mellon University in Pittsburgh. All that's required is to expose the listener to exaggerated, drawn out versions of the distinct sounds. In McClelland's experiments

Japanese adults showed improvement in distinguishing words beginning with *r* and *l* after only three 20- to 25-minute sessions carried out over three days. The training activated a distinct population of neurons accompanied by a gradual strengthening of their connections.

A similar process occurs with facial recognition. By six months of age, infants can perform as well as adults in distinguishing new faces from previously encountered faces. Intriguingly, at nine months, infants do even better than adults at identifying monkey faces they haven't seen before. Neuroscientists discovered this by observing six- and nine month-old infants while they looked at a picture of a new monkey face. The six-month infants looked longer at the new faces, implying that they recognized something new and novel. The nine-month infants, in contrast, looked at the new and the old faces for the same amount of time, a pattern best summed up as, "If you've seen one monkey, you've seen them all."

As with speech recognition, recognizing faces is a two-step process. First comes an early tuning period during which faces of another species are easily distinguished. Then specialization occurs that allows for discernment of faces within a species—among different human or monkey faces, for example. As they mature, infants encounter many human faces but rarely encounter monkey faces. But as with the McClelland findings, adults can recover this talent through training and experience. A person who works in a primate center or a zoo will learn to easily distinguish one monkey face from another. Nor is this skill in cross-species facial recognition confined to monkeys. Farmers can tell one cow from another; horse breeders aren't likely to enter the wrong horse in a race; and when I take my African Grey parrot to the veterinarian for a wing clipping, I can easily identify my bird from other birds in the veterinarian's office that look alike to the non-parrot owner.

The infant language and facial recognition research provide insight into what is happening in the brain as it matures. As the brain becomes more specialized (e.g., recognizing human faces on the basis of experience), neurons previously available for the recognition of a broad array of faces are recruited into the circuits devoted to recognizing only human faces. Next, these circuits are reinforced through repetition so that recognition among human faces becomes finely tuned. The same process occurs with speech: Neurons previously devoted to speech recognition for multiple languages are recruited for the exclusive recognition of native speech sounds.

This general principle of reinforcement through repetition has practical applications for everyday life. For instance, if you want to learn a new skill or make use of new knowledge, you must change your brain. You must engage in repetitive exercises that set up the relevant circuits and sharpen their expression. This holds true whatever your goal and whatever degree of mastery you seek.

chapter 2

Genius and Superior Performance: Are We All Capable?

People with extraordinary abilities, it's turning out, have learned to use their brains differently from the average person. Take chess grand masters, for instance.

Measurements of brain activity during a chess game reveal that grand masters activate their frontal and parietal cortices, two brain areas known to be involved in long-term memory. Skilled amateurs, on the other hand, activate their medial temporal lobes, areas involved in coding new information. This preferential activation of the frontal and parietal cortices by the grand masters—who have memorized thousands of moves over their lifetimes—suggests that they rely on long-term memory both to recognize positions and problems and to retrieve the solutions. Use of the medial temporal lobe by the amateurs, in contrast, suggests a less-effective strategy of analyzing on a case-by-case basis the best response to moves and positions not previously encountered.

The chess master's expertise, in other words, involves storage in the frontal lobes of vast amounts of chess information, a process that takes a long time and a lot of hard work. Grand masters typically spend a minimum of 10 years amassing in their brains an estimated 100,000 or more pieces of chess information (opening gambits, strategies, end-games, etc.). Thanks to this rich memory store, the grand master can quickly assess the advisability and potential consequences—many plays ahead—of a specific move. This ability is one of the reasons a grand master's performance against a good amateur always improves under time pressure, when moves must be made rapidly. "They don't have to think; they are recognizing patterns," according to Ognjen Amidzic, a researcher who has carried out studies of master chess players.

So, would simply learning more moves and deepening your knowledge of chess turn an amateur player like yourself into a grand master? Not necessarily. The genius of the grand master doesn't just depend on the amount of chess information stored in long-term memory, but also on the organization of those memories and how efficiently they can be retrieved. In essence, the grand master must put long-term memory to short-term use.

Geniuses in fields other than chess share a similar talent for storing vast amounts of information in long-term memory and then retrieving that information as circumstances demand. For instance, PET scan studies of geniuses and prodigies in varied pursuits reveal increased brain activity in areas known to be important in the formation of long-term memory. One study compared the PET scans of a German math prodigy, Rudiger Gamm, to scans of people with no special calculating skill.

When doing mental arithmetic, Gamm's brain—but not the brains of the controls—showed activity in areas involved in long-term memory. It's speculated that Gamm uses his long-term memory to store the working results that he needs to complete his calculations. Thus, when rapidly performing mental calculations, he is less likely to lose his place. And if memory is like a notepad—an analogy frequently employed by scientists who study geniuses—then the memory of a Rudiger Gamm is like a library of notepads. Yet this conclusion leaves unanswered a very important question: Is the formation of larger-than-normal long-term memories a genetic trait? Or does it depend on one's individual effort? Important implications ensue from the answer to this question.

What Is a Person's True Potential?

If genius is entirely genetic, then pessimism about achieving eminence would seem to be justified: if genes are so important, most people are likely to remain permanently mired in mediocrity. If, on the other hand, individual effort can lead to enhancement of the brain's structure and function, resulting in the formation of the prodigious long-term memories required for superior performance, then all of us may be capable—if not of genius—of at

least achieving levels of performance that will separate us from
the vast majority of our competitors.

I put this question of inheritance versus development to Anders Ericsson, a psychologist from Florida State University in
Tallahassee who has spent the past 20 years studying geniuses,
prodigies, and other superior performers in fields as diverse as
sports, the arts, and entertainment.

Ericsson is firmly convinced that there are no special inherited qualities that distinguish persons with expert abilities. The
key ingredient turns out to be the willingness to "stretch yourself
to the limit and increase your control over your performance,"
he says. He points to a study he carried out at the highly regarded
Music Academy of West Berlin. "Superior" students, judged by
their teachers as most likely to go on to concert careers, put in an
average of 24 practice hours per week. "Good" students, thought
more likely to end up as teachers than performers, practiced an
average of only 9 hours per week. By the age of 20, the probable
teachers had put in an estimated 4,000 hours of practice, while
the future performers had wracked up an estimated 10,000 hours.

Ericsson found a similar pattern of intense solitary, deliberate
practice among superior performing athletes, chess players, and
mathematicians.

He concludes, "For the superior performer the goal isn't just
repeating the same thing again and again but achieving higher
levels of control over every aspect of their performance. That's
why they don't find practice boring. Each practice session they
are working on doing something better than they did the last
time."

In order to achieve superior performance in a chosen field, the expert must counteract the natural impulse to gain an automated performance as soon as possible. For example, when we first learn to drive, we concentrate, focus our attention, and activate our frontal lobes in order to learn as quickly as possible those maneuvers involving our hands and feet that will enable us to drive efficiently and safely. We approach driving as only a means to an end; once an acceptable performance is achieved (sufficient to pass a driving test), most of us give up on further deliberate efforts at enhancing our driving performance.

Contrast this with the aims and methods of a Grand Prix or other professional racecar driver. He or she must continually exert full concentration and focus during training and competition. Each racecourse is different and calls for different approaches aimed at achieving the fastest speed possible without increasing the risk of an accident. As a result, the professional racecar driver is never completely satisfied with a driving performance but is constantly making adjustments aimed at improving separate components of the driving experience.

For example, in order to guarantee top performance the driver 's frontal lobes must take into account different road conditions (rain or fog, the ambient temperature, the traction that is possible on a particular road surface). The frontal lobes then evaluate, compare, and analyze each variable in order to tally instantaneously a cost-benefit analysis. The professional driver knows that he or she must drive faster than any of the competition but not so fast that the race is lost because of injury or mechanical breakdown.

In contrast, the everyday highway driver's frontal lobes approach the driving experience according to a different program. She isn't in competition with other drivers, needs to emphasize safety over speed, and, if conditions become threatening enough, will pull over and await more friendly driving conditions. The average driver has to do this because she lacks the competitive driver's ability to respond to the many separate components required when handling a car under challenging road conditions.

In summary, the expert must consciously counteract the tendency to prematurely automate the driving experience. While the average driver wants to learn as quickly as possible in order to simultaneously engage in other activities, such as conversing with passengers, the racecar driver must remain involved 100 percent in the driving experience. Further, he or she must develop the capacity to break that experience into multiple components and work on each of these components separately.

"Individuals who perform at higher levels utilize specific kinds of memory processes," according to Ericsson. "They have acquired refined mental representations to maintain access to relevant information and support more extensive flexible reasoning about encountered tasks or situations. Better performers are able to rapidly encode, store, and manipulate information. None of these adjustments would be possible if expert performance were fully automated."

The famous cellist Pablo Casals once summarized this approach to playing even very familiar pieces of music. "To play it perfectly every piece should be studied with the constant idea of improvement in mind, and it is seldom, working in this way, that I do not find that I can improve some one or another detail."

Other superior performers have taken a similar approach to their music. Writing in 1913, the concert pianist F. B. Busoni referred to a perfection of performance that is perpetually sought but is never achieved. "I never neglect an opportunity to improve no matter how perfect a previous interpretation may have seemed to me. In fact, I often go directly home from a concert and practice for hours upon the very pieces that I have been playing because during the concert certain new ideas came to me."

"Expert performers deliberately acquire and refine cognitive mechanisms to enhance their control and monitoring of their performance," says Ericsson. "By comparing their performance to that of more proficient individuals, they can identify differences and gradually reduce them through extended deliberate practice."

The same process holds true for sports. Ericsson points to a remark by one of the most accomplished golfers of the twentieth century, Ben Hogan: "While I am practicing I am also trying to develop my powers of concentration. I never just walk up and hit the ball. I decide in advance how I want to hit and where I want it to go . . . Adopt the habit of concentrating to the exclusion of everything else while you are at the practice tee and you will find that you are automatically following the same routine while playing a round in competition."

Other expert golfers agree that the key component of deliberate practice is to work on those aspects of your performance that you don't do well or those situations that involve an extra challenge. Golf coach Jim McLean makes concentration the centerpiece of successful golf. "Practice only as long as you can concentrate. Stop when you . . . lose focus. Short, focused practice sessions are often the most productive."

"It is only human nature to want to practice what you can already do well, since it's a hell of a lot less work and a hell of a lot more fun. Sad to say, though, it doesn't do a lot to lower your handicap," according to Sam Snead, voted the fourth-best golfer of the twentieth century. "I know it's a lot more fun to stand on the practice tee and rip your driver than it is to chip and pitch, or practice sand shots with sand flying back in your face, but it all comes back to the question of how much you're willing to pay for success."

The Expert's Brain Engages Differently from the Amateur's

So what's going on within the brain? The expert—as opposed to the amateur—golfer, by mentally attending to extremely subtle aspects of his performance during practice sessions, has successfully transferred this knowledge into working memory within his frontal lobes. Later, when under pressure, his brain concentrates on one or more components of his learned procedural skills. For him, task and not ego remain at the forefront of mental activity. This focused approach immunizes the expert player against "choking" under the intense pressure of competitive play.

The amateur, on the other hand, falls victim to a well-established relationship involving arousal, attention, and performance. In a state of heightened anxiety and/or arousal the amateur turns his attention inward and becomes self-focused rather than task-focused. This disrupts the execution of previously learned subunits of performance, resulting in choking under pressure. As Thomas Carr, a psychologist at Michigan State University in East Lansing, explains, "Pressure-induced attention to well-learned components of a proceduralized skill disrupts execution."

Carr is an expert on the factors that govern choking. In essence, he's discovered that it's best to not let self-consciousness get in the way when carrying out a well-learned performance—an idea often expressed by performers who have learned to relax under pressure rather than become self-conscious.

"My approach to performing is: The time to be careful is when you prepare. But, in performing, I say, take a chance, go for it," says concert violinist Nadja Salerno-Sonnenberg.

Theories of "flow" and "inner tennis" are based on enacting similar principles: Don't let self-consciousness and self-evaluation of personal performance interfere with *just doing it*—bringing off a winning performance.

In terms of brain performance, "just doing it" involves the smooth non–self-conscious transfer of learned actions from working memory, stored in the frontal lobes, to the premotor and motor areas that transform the working memory into those effective, winning plays that result from thousands of hours of practice on the part of the performer. Recall from chapter 1 that this is essentially the process Leslie Ungerleider demonstrated with fMRI scans of volunteers who learned a simple sequence of finger movements. Thus, both comparatively straightforward activities and highly sophisticated ones, like learning to become a star athlete or musician, involve taking advantage of the brain's plasticity in order to set up the necessary programs for excellence.

The 10-Year Rule

On the basis of research, Ericsson has formulated what he calls the 10-year rule: "The highest levels of performance and achievement appear to require at least around 10 years of intense prior preparation," he says. Moreover, he believes that anyone who puts in the necessary time can eventually achieve prodigy-level performance.

French psychologist Alfred Binet provided a famous demonstration that anybody can become a prodigy. He pitted two math prodigies against three university students and four cashiers at the Paris department store, Bon Marche. While the prodigies easily outperformed the students, they lagged behind the performance of the cashiers on such challenges as multiplying as rapidly as possible 7,286 by 5,397. How was this possible?

The cashiers averaged 14 years of job experience, which in those days included multiplying a wide range of quantities (lengths of cloth, weights of food, numbers of items) by a wide range of unit prices. On the basis of their daily experience over many years in performing these complicated calculations, the cashiers were able to outperform two prodigies who made their living giving public demonstrations of their calculating prowess.

Both the cashiers and Rudiger Gamm provide partial confirmation of Ericsson's point: Practice hard enough and long enough and you too can become a prodigy. Gamm didn't display any signs of enhanced mathematical ability until, starting at age 20, he started devoting at least four hours a day to performing calculations and memorizing theorems, formulas, and other mathematical facts. Further, his expertise is limited; he doesn't perform with exceptional aptitude in any area other than mathematics.

But Gamm also differs from Ericsson's profile: He reached his prodigious powers in less than 10 years. And many other examples can be sited of people who required less than a decade to achieve superior, even prodigy-level, performance. Mozart, for example, wrote his first composition at age five and the next year was publicly performing on violin and piano.

Ericsson has turned up a possible explanation why some people can achieve genius-level performances in less than 10 years. "The key is to increase one's control over each component of the performance." He came to this conclusion after studying the practice habits of the 10 most highly regarded golfers of the twentieth century. Each exhibited an intense concentration, leading to heightened awareness of each of the many components of their actions coupled with an ability to adapt in ways that led to higher levels of control. In order to alter subtle aspects of their performance, many of them studied videotapes of past tournaments.

While motor and premotor programs can be established at any age, training during childhood and early stages of development yield especially powerful results. For example—and contrary to popular opinion—perfect pitch (the ability to name individual tones) isn't necessarily inherited but can be acquired by average children between three and five years of age if given appropriate training. The development of perfect pitch is accompanied by structural brain changes that fail to appear in the brains of musicians lacking that sensibility. Thanks to plasticity, brain changes also occur as musical talent matures. Further, the resulting changes vary depending on the instrument selected by the performer and the performer's practice patterns. For musicians who play stringed instruments, to take one example, the size and elaboration of cortical area in the brain given over to the fingers, especially the little finger of the left hand, correlates with the age that the person begins musical training.

Future research with fMRI and other technologies are likely to reveal individual "signatures:" patterns of brain activity that vary from person to person and perhaps even from one musical composition to another. The discovery of such individualized patterns will enable performers and their teachers to correlate enhanced musical skills with changes in brain organization and function.

For instance, the brain of a musician who achieves mastery over her instrument frequently undergoes reorganization so that the musical center—the area activated when playing or listening to music—shifts from the right hemisphere (the usual site in amateur musicians) to the left hemisphere. By measuring this change, neuroscientists can presently divide professional-level performers from lesser-skilled ones simply by looking at their respective PET scans. As a further development of this technology, brain scans coupled with electrical measurements will one day provide a flow diagram of the brain's activation patterns while a musician plays a specific composition.

And such fine-grained analyses won't be confined to music. As the technology becomes smaller and more portable, neuroscientists will be able to see what's going on in the brains of expert performers in many fields as they practice their crafts.

A Rage to Master

So, are the achievements of superior performers based entirely on intensely focused training combined with an iron resolve? Or does genetic inheritance play at least some small part as well? As with many either-or questions, this one, I believe, is unanswerable. For one thing, if inheritance plays a part, it's not clear exactly what's inherited. Perhaps the inheritance isn't a particular talent but rather the willingness on the part of the superior performer to put in long hours of solitary practice. In other words, perhaps the inheritance involves what one psychologist has called "a rage to master:" the willingness of the genius, prodigy, or superior performer to devote almost all waking hours to mastering his or her chosen field of endeavor. According to this theory, the chess master would employ the same dedication and intensity for tennis or golf if he dedicated his efforts toward mastering one of these sports instead of chess. Would he be as successful? Personally, I doubt it. With the exception of rare geniuses throughout history (Newton, Descartes), very few people attain international levels of achievement in more than one area.

In short, there is no way of definitively deciding the nature/nurture problem. Genes must always exert their influence within *some* environment. And no matter how powerful the genetic inheritance, the environment must be conducive to the development of a particular talent. A musical talent—even the makings of a musical prodigy—can wither unless the would-be prodigy finds himself among people who appreciate music and help him further his talents.

On the other side of the question, can intensely focused practice in a particular area enable a person with only average talent to achieve the highest levels of performance? After studying and thinking about this question for some time and interviewing experts in the field, such as Ericsson, I find such a proposal intriguing and hopeful but not entirely persuasive. While focused, intensive, deliberate practice can lead to superior performance, I'm not entirely convinced there isn't a small, but nonetheless extremely important, hereditary component as well.

Perhaps Thomas Edison said it best: Genius is 99 percent perspiration and 1 percent inspiration. While intense, focused concentration and practice can enable most of us to achieve a superior performance *if we're willing to put in the necessary work*, there's still that small but important 1 percent that at the highest levels of performance can make the difference between being the best and being *almost* the best.

Brain science in the twenty-first century isn't likely to entirely dispel or explain the mystery of genius or superpowered performance. But whatever the exact proportions of natural talent and hard work, superior performance is ultimately based on the brain's inherent plasticity. This makes for a rule that is both simple and easily applied: Select a field of endeavor that appeals to you and then work with sufficient intensity to bring about major reorganizations in your brain's circuitry.

The Little Gray Cells

Along with insights into genius and superior performance, brain research is also telling us a lot about human intelligence. Until now scientists and educators have measured intelligence by means of so-called intelligence tests.

From our earliest years our lives are greatly affected by how well we perform on intelligence tests. Indeed, these tests play a major role in determining the schools we attend, the people we meet, the friendships we establish, and the spouses we end up with. But what do intelligence tests actually measure? Only the ability to perform well on intelligence tests—as some experts have only half-facetiously suggested—or do they measure something more general, some factor, that distinguishes the person of "high" intelligence from his or her more modestly endowed neighbors?

Numerous studies have shown that if a battery of mental tests is given to a large group of people, a person doing well on one test is also likely to do well on others. Psychologists refer to this "general" mental ability as the g factor. References to a g factor underlie the claim that intelligence differs quantitatively from one person to another. Just as one person can be shown to be "richer" than another by reference to bank accounts, securities, and so on, one person can be confidently declared "smarter" than another on the basis of I.Q. test results and g factors—or so it is claimed.

Until recently the g factor was only a psychological construct—an interesting but rather shadowy attribute that couldn't be physically pinned down to any specific area in the brain. But now all of that is changing as PET scans pinpoint brain areas and brain organizational patterns associated with superior performance on tests of general intelligence.

One of my favorite slides that I show during my brain lectures depicts the PET scan activity of two students measured while they are taking the Raven Progressive Matrices, a test of reasoning and intelligence. The PET scan of one brain shows a lot of activity (an increase of red and orange in the color-coded image), while the other scan shows far less activity (green and blue colors corresponding to decreased activation). Take a guess which of the two PET scans corresponds to the person who scored highest on the Raven. Intuitively, you might correlate superior intelligence and a higher Raven performance with increased PET scan activity. But your intuition would be incorrect. The student with the higher score has the less active PET scan compared to the second student who scored lower on the Raven. What's going on here?

According to Michael Posner and Marcus Raichle, authors of *Images of Mind*, the brain doesn't have to work as hard when engaged in something it finds easy. Thus, the person who scores high in the Raven test doesn't have to strain as much to get the correct answers as does the person scoring lower.

"The efficiency of a computation improves with repetition, and as a result, the overall activity that accompanies the computation is reduced," write Posner and Raichle. "Blood flow is less, electrical activity is reduced, and there is less interference between the repeated computation and any other activity."

Other PET scan work on intelligence supports Posner's and Raichle's point.

As described in an already groundbreaking 2001 paper with the imposing title "A Neural Basis for Intelligence," John Duncan of the Medical Research Council Cognition and Brain Sciences Unit in Cambridge, United Kingdom, carried out PET scans on 13 men and women while they took an I.Q. test. The test contained both verbal and spatial I.Q. style questions.

Duncan found that the solving of such problems isn't distributed throughout the brain—as might be expected in a test for overall intelligence—but is instead concentrated in the lateral prefrontal cortex on each side of the brain. As the name implies, the lateral prefrontal cortex is located toward the front and outer sides of the brain.

He also found that verbal tests activated the left prefrontal cortex (no surprise here since language is processed in the left hemisphere), while spatial problems turned up the lateral prefrontal cortex in both hemispheres. What would account for this preeminent importance of the prefrontal cortex in intelligence?

For one thing, this well-connected brain area transmits and receives information from multiple brain sites. Functionally, it helps us keep several things in mind at once, solve problems, come up with novel ideas, and screen out irrelevant and distracting information. And as mentioned earlier, the frontal lobes are also important in formulating the motor programs required for superior athletic and musical performance. Because of its role in such important functions, the prefrontal cortex would seem a "natural" as a general center for intelligence.

General intelligence (*g*) is also closely linked to the amount of gray matter in the frontal lobes. (Gray matter is made up of brain cells; white matter consists of large tracts of connections uniting one brain cell with another.) Intelligence, it's turning out, has more to do with the number of brain cells in the frontal lobes than with the density of fiber tracts connecting them. On average, people with more gray matter in their frontal lobes tend to perform better on intelligence tests.

The notion that intelligence is related to the amount of gray matter is not new; even our everyday speech is replete with references. We describe a dull person as lacking in gray matter. Hercule Poirot, the Belgian detective immortalized in the novels of Agatha Christie, spoke of using the "little gray cells" to help him identify murderers and their motives.

Not only is the gray matter increased among people scoring well on I.Q. tests, but that increase is also strongly dependent on heredity. The gray matter in identical twins (monozygotic) who share the same genes is more closely matched than in fraternal twins (dizygotic) where the genetic inheritance varies. Identical twins also tend to generally score about the same on I.Q. testing. Further, those twins with the greater gray-matter volume in the frontal region tended to score highest on I.Q. testing.

Do these findings imply that intelligence is pretty much set at birth? That we inherit our level of intelligence and that there isn't much that any of us can do about it? Happily (for those of us who aren't born geniuses), intelligence is plastic and modifiable. All of our experiences result in the formation of neuronal circuits. The richer, more varied, and more challenging the experiences, the more elaborate the neuronal circuits. Of course there are limits: a person testing "average" on intelligence tests can't turn himself into an Einstein simply by signing up for a physics class. Nor can the I.Q. test results be dramatically increased, especially when it comes to abstract reasoning and some other cognitive processes. In fact, performances in these areas tend to decrease slightly with advancing years as a result of a decrease in the functioning of the frontal lobes. But the brain's capacity for change—its plasticity—remains vigorous throughout our lives.

While imaging can help in pinning down important sites in the brain associated with intelligence, I don't think it's likely that traditional methods of defining intelligence are going to be entirely replaced by brain scanning. And even in situations where differences are noted, such as the increase in frontal gray matter with higher intelligence, we encounter a chicken-and-egg problem. As behavioral neuroscientist Robert Plomin puts it,

"People with a stronger motivation might exercise their brains harder and thereby develop a high density of neurons in their frontal cortex."

So, despite the perennial fascination with defining and localizing intelligence, efforts to do so are likely to remain stymied by two major impediments. No agreement exists about what intelligence is, or even what available intelligence tests are actually measuring. Further, as pointed out by psychologist Robert J. Sternberg, "There is a great deal more to everyday living than what intelligence tests measure, or even what can reasonably be called intelligence. The tests fail to take into account such important aspects of functioning as motivation, social skills, persistence in the face of adversity, and the ability to set and achieve reasonable goals."

It's best to think of intelligence, as with genius and superior performance, as partly genetic and partly environmental—and therefore modifiable. Further, since many genes are involved in shaping the structure and functioning of the brain, the existence of a "smart" gene is highly unlikely. Rather, many genes contribute to intelligence and determine the type of intelligence displayed by each of us (the intelligence of the artist compared to that of the businessman, for example). Moreover, thanks to our own efforts we can modify our brain's structure and improve aspects of our intelligence, even if we don't raise our overall I.Q. on standard tests. Rather than being buffeted by the forces of determinism, we retain—thanks to the brain's plasticity and our own efforts—the control necessary to enable our brain to function at our personal best.

Birds of a Feather

Thirty years ago the Swedish psychologist G. Johansson attached tiny lights to the main joints of several actors and recorded their movements in a dark room. He then showed the moving dots of light to volunteers who rapidly and effortlessly recognized the moving dots as a person walking. In most cases the observers could recognize, from the moving dots of light alone, not only movement but also the person's gender, many of their personality traits and emotions, and complex movements such as dancing.

Contemporary PET scanning provides some insight into how the brain can distinguish biologically based movements, such as walking, from random other movements. In one experiment, a part of the brain called the superior temporal sulcus (STS), which separates portions of the temporal lobes, lights up whenever subjects look at dot patterns made by people moving in the dark; the STS remained unresponsive when subjects looked at sequences of random moving dots.

While motion detection is often of survival value in animals to escape predators, it principally provides social and psychological benefits in our own lives. For instance, within the first few days of life an infant will imitate the actions or facial gestures of adults—sticking out its tongue in imitation when an adult performs the same action. Such behaviors signal the brain's first fledgling efforts at bonding and forming emotional connections with other people. It is also an early example of the infant establishing connections between actions and mental states.

By 18 months of age, an infant will imitate and complete observed human actions, such as pulling apart a toy dumbbell, but will ignore the same actions when carried out by a mechanical device. Scientists speculate that such selective imitation of human action serves as a guide for interpreting other people's behaviors in terms of their intentions and desires.

By imitating another person's behavior we, in a real sense, become like them. In recognition of this, folklore and popular advice are replete with admonitions about the importance of choosing the correct "role models" and hanging out with the "right" friends, since "birds of a feather stick together." And imitation, we are told, is "the sincerest form of flattery."

"By simulating another person's actions and mapping them onto stored representations of our own motor commands and their consequences, it might be possible to estimate the observed person's internal states, which cannot be read directly from their movements," write neuroscientists Sarah-Jayne Blakemore and Jean Decety. They add, "This simulating system could also provide information from which predictions about the person's future actions could be made."

Although this may sound superficially like "mind reading" based on nonverbal communication, the concept is a good bit subtler. Neuroscientists have recently discovered the existence of "mirror neurons" in the brains of monkeys that discharge both when the monkey performs certain movements and when the animal merely observes another monkey performing the movement. Strong evidence suggests a similar mirroring process in humans—certain nerve cells are activated both during an activity and while observing another person performing the activity.

In one study subjects underwent brain scanning while watching movements that they would later imitate. Brain activation occurred in the same regions involved in actually performing the actions—an indication that during the period of observation the brain areas responsible for planning the movements were already primed for imitation.

Such experimental findings provide, according to Blakemore and Decety, "a new framework for an idea reminiscent of the philosophical concept that we understand other people's minds by covertly simulating their behavior."

A nice demonstration of such simulation is described in the novel *The Private Memoirs and Confessions of a Justified Sinner* by the Scottish novelist James Hogg: "If I contemplate a man's features seriously, mine own gradually assume the very same appearance and character. And what is more, by contemplating a face minutely, I not only attain the same likeness, but, with the likeness, I attain the very same ideas as well as the same mode of arranging them, so that, you see, by looking at a person attentively, I by degrees assume his likeness, and by assuming his likeness I attain to the possession of his most secret thoughts."

Method actors employ similar techniques in order to enter into the mind of their characters. By studying facial expressions, gaits, and other body movements, the actor is able to become what the novelist Joseph Conrad once termed a "secret sharer" of the personality of another person.

The findings of Blakemore and Decety, along with the observations of Hogg, provide sound reasons for choosing good models for one's own behavior. If you want to accomplish something that demands determination and endurance, try to surround yourself with people possessing these qualities. And try to limit the time you spend with people given to pessimism and expressions of futility. Unfortunately, negative emotions exert a more powerful effect in social situations than positive ones, thanks to the phenomena of *emotional contagion.*

I first learned about emotional contagion from psychologist Steven Stosny, who is a nationally recognized expert on road rage. "Anger and resentment are the most contagious of emotions," according to Stosny. "If you are near a resentful or angry person, you are more prone to become resentful or angry yourself. If one driver engages in angry gestures and takes on the facial expressions of hostility, surrounding drivers will unconsciously imitate the behavior—resulting in an escalation of anger and resentment in all of the drivers. Added to this, the drivers are now more easily startled as a result of the outpouring of adrenaline accompanying their anger. The result is a temper tantrum that can easily escalate into road rage."

As these examples attest, the brain is a powerful simulating machine designed to detect and respond to a wide range of intentions on the part of other people. Neuroscientists are further exploring how our observations of another person's behavior allow us to infer his or her conscious or even unconscious intentions. They are wagering that as we learn more about this process we will be in a position to monitor our own behavior and insure that it is not simply a response to the behavior of others.

chapter 3

Attention Deficit: The Brain Syndrome of Our Era

The plasticity of our brains, besides responding to the people and training to which we expose it, also responds, for good or for bad, to the technology all around us: television, movies, cell phones, e-mail, laptop computers, and the Internet. And by responding, I mean that our brain literally changes its organization and functioning to accommodate the abundance of stimulation forced on it by the modern world.

This technologically driven change in the brain is the biggest modification in the last 200,000 years (when the brain volume of *Homo sapiens* reached the modern level). But while biological and social factors, such as tool use, group hunting, and language, drove earlier brain changes, exposure to technology seems to be spurring the current alteration. One consequence of this change is that we face constant challenges to our ability to focus our attention.

For example, I was recently watching a televised interview with Laura Bush. While the interview progressed, the bottom of the screen was active with a "crawler" composed of a line of moving type that provided information on other news items.

Until recently, crawlers were used to provide early warning signs for hurricanes, tornados, and other impending threats. Because of their rarity and implied seriousness, crawlers grabbed our immediate attention no matter how engrossed we were in the television program playing out before our eyes. Crawlers, in short, were intended to capture our attention and forewarn us of the possible need for prompt action. But now, the crawler has become ubiquitous, forcing an ongoing split in our attention, a constant state of distraction and divided focus.

During the First Lady's interview I found my attention shifting back and forth from her remarks to the active stream of short phrases running below. From the crawler I learned that National Airport was expected to be opened in two days since its closure in the wake of the September 11 terrorist attacks; that this season's Super Bowl would be played in New Orleans one week later than usual; and that a home run record was about to be broken by Barry Bonds.

Despite my best efforts to concentrate on Laura Bush's words, I kept looking down at the crawler to find out what else might be happening that was perhaps even more interesting. As a result, at several points I lost the thread of the conversation between the First Lady and the interviewer. Usually, I missed the question and was therefore forced to remain in the dark during the first sentence or so of her response.

On other occasions I've watched split-screen interviews, with each half of the screen displaying images or text of the topic under discussion while the crawler continues with short snippets about subjects totally divorced from the interview and accompanying video or text. In these instances I am being asked to split my visual attention into three components.

One can readily imagine future developments when attention must be divided into four or more components—perhaps an interview done entirely in the form of a voice-over, with the split-screen video illustrating two subjects unrelated to the subject of the interview and accompanied all the while by a crawler at the bottom of the screen dealing with a fourth topic.

Yet we shouldn't think of such developments as unanticipated or surprising. In 1916, prophets of the Futurist Cinema lauded "cinematic simultaneity and interpenetrations of different times and places" and predicted "we shall project two or three different visual episodes at the same time, one next to the other." Yesterday's predictions have become today's reality. And in the course of that makeover we have become more frenetic, more distracted, more fragmented—in a word, more *hyperactive*.

How Many Ways
Can Our Attention Be Divided?

Divisions of attention aren't new, of course. People have always been required to do more than one thing at a time or think of more than one thing at a time. But even when engaged in what we now call multitasking, most people maintained a strong sense of unity: They remained fully grounded in terms of what they were doing. Today the sense of unity has been replaced, I believe, by feelings of distraction and difficulty maintaining focus and attention. On a daily basis I encounter otherwise normal people in my neuropsychiatric practice who experience difficulty concentrating. "I no sooner begin thinking of one thing than my mind starts to wander off to another subject and before I know it I'm thinking of yet a third subject," is a typical complaint.

Certainly part of this shift from focus to distraction arises from the many and varied roles we all must now fulfill. But I think the process of personal *dis*-integration is also furthered by our constant exposure to the media, principally television. When watching TV, many of us now routinely flit from one program to another as quickly as our thumb can strike the remote control button. We watch a story for a few minutes and then switch over to a basketball game until we become bored with that, and then move on to Animal Planet. Feeling restless, we may then pick up the phone and talk to a co-worker about topics likely to come up at tomorrow's meeting while simultaneously directing our attention to a weather report on TV or flipping through our mail.

"The demands upon the human brain right now are increasing," according to Todd E. Feinberg, a neurologist at Beth Israel Medical Center in New York City. "For all we know, we're selecting for the capacity to multitask."

Feinberg's comment about "selecting" gets to the meat of the issue. At any given time evolution selects for adaptation and fitness to prevailing environmental conditions. And today the environment demands the capacity to do more than one thing at a time, divide one's attention, and juggle competing, often conflicting, interests. Adolescents have grown up in just such an environment. As a result, some of them can function reasonably efficiently under conditions of distraction. But this ability to multitask often comes at a price—Attention Deficit Disorder (ADD) or Attention Deficit Hyperactivity Disorder (ADHD).

Perhaps the best intuitive understanding of ADD/ADHD comes from the French philosopher Blaise Pascal who said, "Most of the evils in life arise from a man's being unable to sit still in a room." The fourth edition of the *Diagnostic and Statistical Manual of Mental Disorders* (DSM-IV) provides a more contemporary definition. Although ADD/ADHD affects adults as well as children, the DSM-IV describes symptoms as they affect three categories in children: motor control, impulsivity, and difficulties with organization and focus.

The motor patterns include:

(a) often fidgets with hands or feet or squirms in seat
(b) often leaves seat in classroom or in other situations in which remaining seated is expected

(c) is often "on the go" or often acts as if driven by a motor

(d) often runs about or expresses a subjective feeling of rest-lessness

(e) often has difficulty playing or engaging in leisure activi-ties quietly

(f) often talks excessively

The impulsive difficulties include:

(a) often experiences difficulty awaiting turn

(b) interrupts or intrudes on others (e.g., butts into conversa-tion or games)

(c) often blurts out answers before questions have been com-pleted

To earn the diagnosis of the "inattentive subtype" of ADD/ADHD, the child or adolescent shows any six of the fol-lowing symptoms:

(a) often does not follow through on instructions and fails to finish schoolwork, chores, or duties in the workplace

(b) often fails to give close attention to details or makes care-less mistakes in schoolwork, work, or other activities

(c) often has difficulty sustaining attention in tasks or play ac-tivities

(d) often does not seem to listen when spoken to directly

(e) often avoids, dislikes, or is reluctant to engage in tasks that require sustained mental effort (such as schoolwork or homework)

(f) often loses things necessary for tasks or activities

(g) is often easily distracted by extraneous stimuli

(h) is often forgetful in daily activities

For years doctors assured the parents of an ADD/ADHD child that the condition would disappear as their child grew older. But such reassurances have turned out to be overly optimistic. In the majority of cases, ADD/ADHD continues into adulthood, although the symptoms change.

In their best-selling book, *Driven to Distraction*, psychiatrists Edward Hallowell and John Ratey developed a list of criteria for the diagnosis of Adult Attention Deficit Disorder. Among the most common manifestations are:

1. A sense of underachievement, of not meeting one's goals
2. Difficulty getting organized
3. Chronic procrastination or trouble getting started
4. Many projects going simultaneously; trouble with follow-through
5. A tendency to say whatever comes to mind without necessarily considering the timing or appropriateness of the remark
6. A frequent search for high stimulation
7. Intolerance of boredom
8. Easy distractibility, trouble in focusing attention, a tendency to tune out or drift away in the middle of a page or conversation
9. Impatient; low frustration tolerance
10. A sense of insecurity

Other experts on adult attention disorder would add:

11. Low self-esteem and
12. Emotional lability: sudden and sometimes dramatic mood shifts

A Distinctive Type of Brain Organization

In many instances childhood and adult ADD/ADHD is inherited. Typically, the parents of a child diagnosed with the disorder will be found upon interview to exhibit many of the criteria for adult ADD/ADHD. But many cases of ADD/ADHD in both children and adults occur without any hereditary disposition, suggesting the probability of culturally induced ADD/ADHD.

As a result of increasing demands on our attention and focus, our brains try to adapt by rapidly shifting attention from one activity to another—a strategy that is now almost a requirement for survival. As a consequence, attention deficit disorder is becoming epidemic in both children and adults. This is unlikely to turn out to be a temporary condition. Indeed, some forms of ADD/ADHD have entered the mainstream of acceptable behavior. Many personality characteristics we formerly labeled as dysfunctional, such as hyperactivity, impulsiveness, and easy distractibility, are now almost the norm.

"With so many distracted people running around, we could be becoming the first society with Attention Deficit Disorder," writes Evan Schartz, a cyberspace critic in *Wired* magazine. In Schartz's opinion, ADHD may be "the official brain syndrome of the information age."

"Civilization is revving itself into a pathologically short attention span. The trend might be coming from the acceleration of technology, the short-horizon perspective of market-driven economies, the next-election perspective of democracies, or the distractions of personal multitasking. All are on the increase," according to Stewart Brand, a noted commentator on technology and social change.

As ADD expert Paul Wender puts it: "The attention span of the average adult is greatly exaggerated."

"It's important to note that neuroscientists and experts within the field are increasingly dissatisfied with ADHD being called a disorder," according to Sam Horn, author of *Conzentrate: Get Focused and Pay Attention—When Life Is Filled with Pressures, Distractions, and Multiple Priorities*, which lists forces in the modern world that "induce" ADD/ADHD. "They prefer to see ADHD as a distinctive type of brain organization."

Such an attitude change toward ADD/ADHD carries practical implications. When creating an optimum environment for learning, for instance, Horn suggests, "blocking out sounds can hurt. Today's younger generation has become accustomed to cacophony. Street sounds, the screeching of brakes, trucks changing gears, and the wails of ambulances are their norm. For these people silence can actually be disconcerting because it's so unusual."

To Horn's list of ADD/ADHD-inducing influences I would add time-compressed speech, which is now routinely used on radio and TV to inject the maximum amount of information per unit of time. As a result we have all become accustomed to rapid-fire motormouth commercials spoken at truly incomprehensible speeds. Think of the last car commercial you saw where all the "fine print" of the latest deal was read with lightning speed, or the pharmaceutical pitch that names a dozen possible side effects in less than five seconds.

"The attitude seems to be one of pushing the limits on the listener as far as 'the market will bear' in terms of degrading the auditory signal and increasing the presentation rate of the spoken programming," according to Brandeis University psychologists Patricia A. Tun and Arthur Wingfield in their paper, "Slow but Sure in an Age of 'Make it Quick.'"

As these psychologists point out, laptop computers, cell phones, e-mail, and fax machines keep us in constant touch with the world while simultaneously exerting tremendous pressures on us to respond quickly and accurately. But speed and accuracy often operate at cross-purposes in the human brain.

In study after study both young and older listeners recall less from materials told to them at a rapid rate. A similar situation exists in the visual sphere. A television viewer's memory for information about the weather is actually poorer after viewing weather segments featuring colored charts and moving graphics than after viewing straight-forward versions of the same information in which the weather is simply described.

As Tun and Wingfield put it: "The clutter, noise, and constant barrage of information that surround us daily contribute to the hectic pace of our modern lives, in which it is often difficult simply to remain mindful in the moment."

No Time to Listen

As the result of our "make it quick" culture, attention deficit is becoming the paradigmatic disorder of our times. Indeed,

ADD/ADHD isn't so much a disorder as it is a cognitive style. In order to be successful in today's workplace you have to incorporate some elements of ADD/ADHD.

You must learn to rapidly process information, function amidst surroundings your parents would have described as "chaotic," always remain prepared to rapidly shift from one activity to another, and redirect your attention among competing tasks without becoming bogged down or losing time. Such facility in rapid information processing requires profound alterations in our brain. And such alterations come at a cost—a devaluation of the depth and quality of our relationships.

For example, a patient of mine who works as a subway driver was once unfortunate enough to witness a man commit suicide by throwing himself in front of her train. Her ensuing anguish and distress convinced her employers that she needed help, and they sent her to me. The hardest part of her ordeal, as she expressed it, was that no one would give her more than a few minutes to tell her story. They either interrupted her or, in her words, "gradually zoned out."

"I can't seem to talk fast enough about what happened to me," she told me. "Nobody has time to listen anymore."

The absence of the "time to listen" isn't simply the result of increased workloads (although this certainly plays a role) but from a reorganization of our brains. Sensory overload is the psychological term for the process, but you don't have to be a psychologist to understand it. Our brain is being forced to manage increasing amounts of information within shorter and shorter time intervals. Since not everyone is capable of making that transition, experiences like my patient's are becoming increasingly common.

"Don't tell me anything that is going to take more than 30 seconds for you to get out," as one of my adult friends with ADD/ADHD told his wife in response to what he considered her rambling. In fact, she was only taking the time required to explain a complicated matter in appropriate detail.

"The blistering pace of life today, driven by technology and the business imperative to improve efficiency, is something to behold," writes David Shenk in his influential book *Data Smog*. "We often feel life going by much, much faster than we wish, as we are carried forward from meeting to meeting, call to call, errand to errand. We have less time to ourselves, and we are expected to improve our performance and output year after year."

Regarding technology's influence on us, Jacques Barzun, in his best-seller, *From Dawn to Decadence*, comments, "The machine makes us its captive servants—by its rhythm, by its convenience, by the cost of stopping it or the drawbacks of not using it. As captives *we come to resemble it in its pace, rigidity, and uniform expectations*" [emphasis added].

Whether you agree that we're beginning to resemble machines, I'm certain you can readily bring to mind examples of the effect of communication technology on identity and behavior. For instance, cinematography provides us with many of our reference points and a vocabulary for describing and even experiencing our personal reality.

While driving to work in the morning we "fast-forward" a half-hour in our mind to the upcoming office meeting. We reenact in our imagination a series of "scenarios" that could potentially take place. A few minutes later, while entering the garage, we experience a "flashback" of the awkward "scene" that took place during last week's meeting and "dub in" a more pleasing "take."

Of course using the vocabulary of the latest technology in conversation isn't new. Soon after their introduction, railways, telegraphs, and telephone switchboards provided useful metaphors for describing everyday experiences: People spoke of someone "telegraphing" their intentions, or of a person being "plugged in" to the latest fashions.

Modern Nerves

In 1891 the Viennese critic Hermann Bahr predicted the arrival of what he called "new human beings," marked by an increased nervous energy. A person with "modern nerves" was "quick-witted, briskly efficient, rigorously scheduled, doing everything on the double," writes social critic Peter Conrad in *Modern Times, Modern Places*.

In the 1920s, indications of modern nerves were illustrated by both the silent films of the age, with their accelerated movement, and the change in drug use at the time, from sedating agents like opium to the newly synthesized cocaine—a shift that replaced languid immobility with frenetic hyperactivity and "mobility mania."

Josef Breuer, who coauthored *Studies on Hysteria* with Sigmund Freud, compared the modern nervous system to a telephone line made up of nerves in "tonic excitation." If the nerves were overburdened with too much "current," he claimed, the result would be sparks, frazzled insulation, scorched filaments, short circuits—in essence, a model for hysteria. The mind was thus a machine and could best be understood through the employment of machine metaphors. Athletes picked up on this theme and aimed at transforming their bodies into fine-tuned organisms capable, like machines, of instant responsiveness. "The neural pathways by which will is translated into physical movement are trained until they react to the slightest impulse," wrote a commentator in the 1920s on the "cult" of sports.

The Changing Rhythm of Life

In 1931 the historian James Truslow Adams commented, "As the number of sensations increase, the time which we have for reacting to and digesting them becomes less . . . the rhythm of our life becomes quicker, the wave lengths . . . of our mental life grow shorter. Such a life tends to become a mere search for more and more exciting sensations, undermining yet more our power of concentration in thought. Relief from fatigue and ennui is sought in mere excitation of our nerves, as in speeding cars or emotional movies."

In the 60 years since Adams's observation, speed has become an integral component of our lives. According to media critic Todd Gitlin, writing in *Media Unlimited*, "Speed is not incidental to the modern world—speed of production, speed of innovation, speed of investment, speed in the pace of life and the movement of images—but its essence. . . . Is speed a means or an end? If a means, it is so pervasive as to *become* an end."

In our contemporary society speed is the standard applied to almost everything that we do. Media, especially television, is the most striking example of this acceleration. "It is the limitless media torrent that sharpens the sense that all of life is jetting forward—or through—some ultimate speed barrier," according to Gitlin. "The most widespread, most consequential speed-up of our time is the onrush in images—the speed at which they zip through the world, the speed at which they give way to more of the same, the tempo at which they move."

In response to this media torrent, the brain has had to make fundamental adjustments. The demarcation between here and elsewhere has become blurred. Thanks to technology, each of us exists simultaneously in not just one *here* but in several. While talking with a friend over coffee we're scanning e-mail on our Palm Pilot. At such times where are we *really*? In such instances no less is involved than a fundamental change in our concept of time and place.

Where Is Where?

"Modernity is about the acceleration of time and the dispersal of places. The past is available for instant recall in the present," according to Peter Conrad. For example, I was recently sitting in a restaurant in Washington, D.C., while watching a soccer match take place several time zones away. During an interruption in play, the screen displayed action from another match played more than a decade ago. The commentator made a brief point about similarities and differences in the two matches and then returned to the action of the ongoing match. During all of this I was participating in a "present" comprised of two different time zones along with a "past" drawn from an event that occurred twelve years earlier. Such an experience is no longer unusual. Technology routinely places us in ambiguous time and place relationships.

As another example, while recently sitting on the beach at South Beach in Miami I was amazed at the number of people talking on cell phones while ostensibly spending the afternoon with the person who accompanied them to the beach. In this situation the *here* is at least partly influenced by the technology of the cell phone that both links (the caller and the unseen person on the other end of the cell phone) and isolates (the caller and the temporarily neglected person lying beside him or her on the beach blanket). Such technologies are forcing our brains to restructure themselves and accommodate to a world of multiple identity and presence.

Intellectually we have always known that the "reality" of the here and now before our eyes is only one among many. But we never directly experienced this multilevel reality until technology made it possible to reach from one end of the world to another and wipe out differences in time, space, and place. Starting with telephones we became able to experience the "reality" of people in widely dispersed areas of the world. With the cell phone, that process has became even more intimate. Time, distance, night, and day—the rules of the natural and physical world—cease to be limiting factors.

And while some of us may celebrate such experiences and thrive on constantly being connected, others feel the sensation of a giant electronic tentacle that will ensnare us at any moment.

My point here isn't to criticize technology but to emphasize the revolution that technology is causing in our brain's functioning. If, for example, through technology, anyone at any given moment is immediately available, "here" and "there" lose their distinctive meanings. We achieve that "acceleration of time and dispersal of places" referred to by Peter Conrad.

And yet there is an ironic paradox in all this: As a result of technological advances we participate in many different and disparate "realities," yet as a result of our attention and focus problems we can't fully participate in them. We can shift back and forth from a phone conversation with someone in Hong Kong and someone directly in front of our eyes. Yet thanks to our sense of distraction we're not fully focused on either of them. What to do?

The Plastic on the Cheese

"If I can only learn to efficiently carry out several things simultaneously then my time pressures will disappear," we tell ourselves. And at first sight multitasking seems a sensible response to our compressed, overly committed schedules. Instead of limiting ourselves to only one activity, why not do several simultaneously? If you owe your mother a phone call, why not make that call while in the kitchen waiting for the spaghetti to come to a boil? And if Mom should call you first, why not talk to her while glancing down at today's crossword puzzle?

Actually, multitasking is not nearly as efficient as most of us have been led to believe. In fact, doing more than one thing at once or switching back and forth from one task to another involves time-consuming alterations in brain processing that reduce our effectiveness at accomplishing either one.

Whenever you attempt to do "two things at once," your attention at any given moment is directed to one or the other activity rather than to both at once. And, most important, these shifts decrease rather than increase your efficiency; they are time and energy depleting.

With each switch in attention, your frontal lobes—the executive control centers toward the front of your brain—must shift goals and activate new rules of operation. Talking on the phone and doing a crossword puzzle activate different parts of the brain, engage different muscles, and induce different sensory experiences.

In addition, the shift from one activity to another can take up to seven-tenths of a second. We know this because of the research of Joshua Rubinstein, a psychologist at the Federal Aviation Administration's William J. Hughes Technical Center in Atlantic City.

Rubinstein and his colleagues studied patterns of time loss

that resulted when volunteers switched from activities of varying complexity and familiarity. Measurements showed that the volunteers lost time during these switches, especially when going from something familiar to something unfamiliar. Further, the time losses increased in direct proportion to the complexity of the tasks. To explain this finding the researchers postulate a "rule-activation" stage, when the prefrontal cortex "disables" or deactivates the rules used for the first activity and then "enables" the rules for the new activity. It's this process of rule deactivation followed by reactivation that takes more than half a second. Under certain circumstances this loss of time due to multitasking can prove not only inefficient but also dangerous.

For example, remember the speculation that cell phone–associated automobile accidents could be eliminated if drivers used hands-free devices? Well, that speculation isn't supported by brain research. The use of cell phones—hands-free or otherwise—divides a driver's attention and increases his or her sense of distraction.

In an important study carried out by psychologist Peter A. Hancock at the University of Central Florida in Orlando and two researchers from the Liberty Mutual Insurance Company, volunteers simulated using a hands-free cell phone while driving. The volunteers were instructed to respond to the ringing of a phone installed on the dashboard of their car. At the instant they heard the ring they had to compare whether the first digit of a number displayed on a computer screen on the dashboard corresponded to the first digit of a number they had previously memorized. If that first digit was the same, the driver was supposed to push a button. In the meantime, they were to obey all traffic rules and, in the test situation, bring the car to a full stop.

While the distracting ring had only a slight effect on the stopping distance of younger drivers (0.61 second rather than 0.5 second), it had a profound effect on the stopping distance of drivers between 55 and 65 years of age: 0.82 second rather than 0.61 second, according to the researchers. Distraction, in other words, reduces efficiency.

In another test of the cost of multitasking, volunteers at the Center for Cognitive Brain Imaging at Carnegie Mellon University in Pittsburgh underwent PET scans while simultaneously listening to sentences and mentally rotating pairs of three-dimensional figures. The researchers found a 29 percent reduction in brain activity generated by mental rotation if the subjects were also listening to the test sentences. This decrease in brain activity was linked to an overall decrease in efficiency: it took them longer to do each task.

A reduction in efficiency was also found when the researchers looked at the effect of mental rotation on reading. They discovered that brain activity generated when reading the sentences decreased by 53 percent if the subjects were also trying to mentally rotate the objects.

A similar loss of efficiency occurs when activities are alternated. For instance, David E. Meyer, a professor of mathematical psychology at the University of Michigan in Ann Arbor, recruited young adults to engage in an experiment where they would rapidly switch between working out math problems and identifying shapes. The volunteers took longer for both tasks, and their accuracy took a nosedive compared to their performance when they focused on each task separately.

"Not only the speed of performance, the accuracy of performance, but what I call the fluency of performance, the gracefulness of their performance, was negatively influenced by the overload of multitasking," according to Meyer.

All of which leads to this simple rule: Despite our subjective feelings to the contrary, actually our brain can work on only one thing at a time. Rather than allowing us to efficiently do two things at the same time, multitasking actually results in inefficient shifts in our attention. In short, the brain is designed to work most efficiently when it works on a single task and for sustained rather than intermittent and alternating periods of time. This doesn't mean that we can't perform a certain amount of multitasking. But we do so at decreased efficiency and accuracy.

But despite neuroscientific evidence to the contrary, we are being made to feel that we *must* multitask in order to keep our head above the rising flood of daily demands. Instead of "Be Here Now," we're encouraged to split our attention into several fragments and convinced that multitasking improves mental efficiency.

Instead, multitasking comes at a cost. And it's true that sometimes the cost is trivial, or even amusing, as with the following experience of a young mother: "I had to get dressed for my daughter's middle school choral program, get another child started on homework, and feed another who had to be ready to be driven to soccer practice. Of course, the phone kept ringing, too. I thought I had everything under control when the complaints about the grilled cheese started. Without getting too angry, I growled for her to just eat it so that I could finish getting dressed. What else could I do in such a rush? My daughter then said, 'I can't eat it, Mom. You left the plastic on the cheese.'"

Other times, the cost of multitasking can be much less amusing. Imagine yourself driving in light traffic on a clear day while chatting on a cell phone with a friend. You're having no problem handling your vehicle and also keeping up your end of the conversation. But over the next five minutes you encounter heavier traffic and the onset of a torrential rainstorm. Your impulse is to end the conversation and pay more attention to the road, but your friend on the other end of the line keeps talking. After all, he isn't encountering the same hazardous conditions from the comfort of his office or home. You continue to talk a bit longer, shifting your attention between your friend's patter and the rapidly deteriorating road conditions. As a result, you fail to notice that the tractor-trailer to your right is starting to slide in your direction. . . Your survivors will never know that your divided attention, with its accompanying decrease in brain efficiency, set you up for that fatal accident.

In essence, the brain has certain limits that we must accept. While it's true that we can train our brain to multitask, our overall performance on each of the tasks is going to be less efficient than if we performed one thing at a time.

Cerebral Geography

Despite the inefficiency of multitasking, the brain is able to deal with more than one thing at a time. If that weren't true, we wouldn't be able to "walk and chew gum at the same time," as a critic once uncharitably described a former U.S. president. The trick is to avoid activities that interrupt the flow of the main activity.

For example, listening to music can actually enhance the efficiency among those who work with their hands. I first learned of this a year or so ago when a draftsman casually mentioned to me that he felt more relaxed and did better work while listening to background music. Many surgeons make similar claims. In a study aimed at testing such claims, researchers hooked up 50 male surgeons between the ages of 31 and 61 to machines that measured blood pressure and pulse. The surgeons then performed mental arithmetic exercises designed to mimic the stress a surgeon would be expected to experience in the operating room. They then repeated the exercise while the surgeons listened to musical selections of their own choosing. The performances improved when the surgeons were listening to the music.

In another study, listening to music enhanced the surgeons' alertness and concentration. What kind of music worked best? Of 50 instrumental tracks selected, 46 were concertos, with Vivaldi's *Four Seasons* as the top pick, followed by Beethoven's Violin Concerto op. 61, Bach's Brandenburg Concertos, and Wagner's "Ride of the Valkyries"—not exactly your standard "easy listening" repertoire.

But easy listening isn't the purpose, according to one of the surgeons interviewed. "In the O.R. it's very busy with lots of things going on, but if you have the music on you can operate. The music isn't a distraction but a way of blocking out all of the other distractions."

Music undoubtedly exerts its positive effects on surgical performance at least partially through its kinesthetic effects, an observation made by Socrates in Plato's *Republic*: "More than anything else rhythm and harmony find their way to the inmost soul and take strongest hold upon it." Thanks to music, the surgeon is more concentrated, alert, technically efficient, and—most important—in the frame of mind most conducive to healing. "Take Puccini's *La Bohème*," says Blake Papsin, an ear, nose, and throat surgeon in Toronto. "It's an absolutely beautiful piece of music that compels the human spirit to perform, to care, to love—and that's what surgery is."

Music and skilled manual activities activate different parts of the brain, so interference and competition are avoided. If the surgeon listened to an audio book instead of a musical composition, however, there would likely be interference. Imagining a scene described by the narrator would interfere with the surgeon's spatial imaging. Listening to the audiobook would activate similar areas of the brain and cause competition between the attention needed to efficiently and accurately operate and to comprehend the images and story in the audiobook. We encounter here an example of the principle of *cerebral geography*: the brain works at its best with the activation of different, rather than identical, brain areas. That's why doodling while talking on the telephone isn't a problem for most people, since speaking and drawing use different brain areas. But writing a thank you note while on the phone results in mental strain because speaking and writing share some of the same brain circuitry.

Thanks to new technology, especially procedures like functional MRI scans, neuroscientists will soon be able to compile lists of activities that can be done simultaneously with a minimal lapse in efficiency or accuracy. But, in general, it's wise to keep this in mind: A penalty is almost always paid when two activities are carried out simultaneously rather than separately.

chapter 4

More Images Than Ever: Is It Destabilizing Our Brains?

So far we have explored how high-performing people work to improve their brains' powers of concentration and focus and how modern life is affecting the demands placed on our brain. But the brain is concerned with more than just logic, efficiency, and outstanding performance. The brain is also the seat of fears, prejudices, and other negative emotions. Recent research points to potentially harmful effects on the brain brought on by exposure to media images of horror and mayhem. Such depictions, it's turning out, activate the same brain areas involved in the feeling and expressing of real-life emotions. In order to understand the relationship between disturbing images and emotional arousal, it's helpful to say a few words about emotional processing within the brain.

All of our emotional expressions and perceptions can be traced to cortical and subcortical neurons collectively referred to as the limbic system: an interconnected network of centers located within and below the cortex, which contributes to the experience and expression of emotion. The most prominent components are pointed out in the illustration at the beginning of this book: the cingulate gyrus, the hippocampus, and the amygdala. This network is especially important in mediating violence and aggression, including violent responses to violations of territory and personal space.

In order to get a feel for the structure of the limbic system, imagine yourself cutting a ripe melon exactly in half. You then separate the two sides and peer at the inner aspect of each half. The most striking structure viewed from this perspective is the corpus callosum—a bridge of fibers connecting the two cerebral hemispheres. The brain tissue forming a rim around the corpus callosum is referred to as the limbic lobe (*limbus* in Latin means fringe) and forms the cortical component of the limbic system.

Notice, in the illustration at the front of this book, the amygdala, a small almond-shaped nucleus (*amygdala* means almond in Latin) at the tip of the temporal lobes on both sides of the brain. This is turning out to be one of the most important structures mediating our emotions. How do we know this? On the basis of some insightful experiments carried out on animals.

An especially revealing experiment from the 1980s by J. L. Downer at the University of London tested how a monkey would react if deprived of the input from its amygdala. To create this situation Downer surgically modified the monkey's brain so that the animal possessed only one functioning amygdala (normally there are two, one on each side). Further, the single functioning amygdala had access only to the visual information provided by the eye on the same side.

Downer discovered that when the monkey looked at the world through the eye connected to the amygdala, the monkey's behavior was typical of a monkey-in-captivity: fearful, aggressive, and marked by attempts at evasion and escape. As Downer described the behavior, "When observers looked into its cage, the animal would bare its teeth and dash forward, jumping at the door and attempting to bite and claw." Then, in the second part of the experiment Downer covered that eye so that the monkey was forced to use the eye disconnected from the amygdala. The animal quieted down and could be easily handled; it even approached Downer and took raisins from his hand.

In short, a monkey deprived of an amygdala loses its natural ability to interpret and appropriately respond to the potential threat of an approaching human. The monkey did respond normally, incidentally, if touched on either side of its body, implying that threat conveyed via the sensation of touch could reach the single amygdala from either side of the body.

Think of the amygdala as a device that, among other activities, evaluates the environment for danger and threat. When it's not working correctly threats may be ignored; hyperactivity of the amygdala, in contrast, may result in excessive fearfulness. Electrical stimulation of the amygdala in animals induces fear responses such as increases in heart rate and breathing, along with the release into the bloodstream of stress hormones. The same response results if an animal undergoes fear conditioning.

In a typical fear conditioning experiment, an animal, usually a rat, is placed in a special cage with a floor designed to deliver an electric shock. At the start of the experiment a sound comes on, followed by a brief, mild shock to the feet. After only a few pairings of the sound and the shock, the rat starts acting frightened when it hears the sound alone. It freezes in place, its fur stands on end, its heart rate and blood pressure rise, and stress hormones pour into the blood stream. But fear conditioning can be blocked if both pathways to the amygdala are destroyed.

A Delicate Balance

Extensively linked to the emotion-mediating limbic system is the prefrontal cortex, which controls fear and aggression. This dual function is useful because fear and aggression are intimately connected. For example, if someone makes a loud startling noise behind your back, you're likely, in your momentary fright, to react angrily against the offender. You might whip around and verbally abuse him or her for their "carelessness" or "stupidity." If sufficiently frightened you might even physically strike out against the offender. That sequence of fright followed by anger originates within the limbic system.

But for most of us, counteracting forces in our personality lead us to weigh the consequences of an overly aggressive response. In most instances we don't hit anybody. We rein in our momentary anger and, a few seconds later, even apologize for any signs of annoyance we may have displayed. The prefrontal cortex is the "coach" that helps us retain control. It counsels restraint, suggests putting everything in context, moving on. Damage to the prefrontal cortex results in a loss of restraint and an inability to inhibit aggression and exercise judgment.

Recent brain research indicates that the prefrontal cortex and the limbic system coexist in a delicate balance. When emotions threaten to get out of control, the prefrontal cortex springs into action and restores the balance. In fact, activity in the prefrontal cortex is an indicator of an imbalance between the limbic system and the prefrontal lobes. While real-life situations are the most frequent disturber of the frontal-limbic balance, research from the National Institutes of Health and elsewhere indicates that images seen on television or in the movies can be equally effective in destabilizing the "thinking" and the "feeling" brain.

Images and Emotional Communication

Images played only a limited role in the communication of emotion prior to the development of photography, the advent of film, and the birth of silent movies in the mid-1890s. Until then all information could only be conveyed via either the printed or spoken word. But photographs and movies changed all that: Images began replacing words as the token of information exchange.

With the development of television, images became even more ubiquitous. It was no longer necessary to go to the movies; television brought images into the home. By 1960 U.S. television viewership hit 50 million, with more households possessing televisions than indoor toilets. Today, television has become so pervasive that most households have two or more television sets of varying sizes distributed throughout the home. Images have become our culture's primary purveyor of information.

According to political scientist Benjamin R. Barber, "Images, reinforced by recorded sound, take the place of words, numbers, and other ciphers with which humans have traditionally communicated."

In addition, the pace at which these images confront us has been enormously accelerated thanks to digitalization and computerization. The pictures and sound associated with an event happening anywhere on the globe can be instantaneously transmitted. And since images—unlike writing or speaking—are immediate and require no thought or analysis, they have overtaken words as the principle means by which knowledge gets communicated. This has important consequences.

According to Barber, this change from print on a page to images on a screen results in a "colossal Re-formation of the human condition. . . . The formats and packaging have changed (television, videos, and computer screens) but the payoff still comes as moving images that pass before human eyes. The abstraction of language is superseded by the literalness of pictures—at a yet to be determined cost to imagination, which languishes as its work is done for it."

It is now possible for us to view and repeat images of upsetting and alarming events as if they are actually happening before us. The horrors of a particular event such as the World Trade Center attack can be relived—often unexpectedly, when we turn on our television and are exposed to the rerun of a clip of a plane hitting a tower. The quality of the images has also improved to the point that depictions of disasters (facial expressions of horror, gory images of death and maiming) are rendered with a verisimilitude that approaches the perceptions of someone actually present. Only recently have neuroscientists gotten a handle on just how destructive such experiences can be.

According to a study that was carried out at the National Institute of Neurological Disorders and Stroke (NINDS) by Marina Nakic, merely viewing violent or aggressive images, such as those routinely encountered on television and in movies, is sufficient to activate the prefrontal cortex. Especially sensitive to activation is the orbitofrontal cortex. This area is in intimate contact with the emotional centers of the amygdala and other limbic system components. Following the orbitofrontal activation, brain circuits are established that encode the images for subsequent replay.

Later, reexposure to the same or similar scenes of violence, horror, or mayhem strengthens the newly established circuits, as well as induces the formation of additional circuits within areas important in emotion. When the emotionally arousing image is reencountered, these circuits are activated and the brain reacts with emotional arousal each time.

Recent neuroscientific research indicates that some people

may be at increased risk for psychological harm when looking at disturbing images, according to Richard Davidson, a professor of psychology and psychiatry at the University of Wisconsin-Madison. In his laboratory, Davidson uses PET and fMRI imaging to compare the brains of people who always seem to be overly sad and pessimistic with the brains of those who remain upbeat even under the most trying of circumstances.

"We have discovered that there are differences in certain circuits in the brain that differentiate between these positive, happy individuals compared to those who show more vulnerability in response to the emotional events in their lives," he says.

Davidson has observed increased PET and fMRI activation within the left prefrontal cortex among individuals with a positive, happy outlook. Such people also show inhibition in the amygdala, which I have already shown to be important in governing emotional response. Among unhappy, vulnerable people just the opposite occurs: increased amygdalar activation along with an increase in activity of the right, not the left, prefrontal cortex. These differences have practical implications when it comes to watching video images.

Among adults, Davidson has found that certain patterns of activity in the lobes of the prefrontal cortex predict how a person is likely to respond to films of positive or negative scenarios—a party versus an auto accident, for example. Those with greater activity on the right side are more likely to experience distressing emotions when viewing a negative scenario. Those with activity on the left side do not respond as traumatically.

In other words, when exposed to video images of horror and

mayhem not everyone is going to be affected in the same way. The problem is that at the moment there isn't any way of predicting—short of brain scanning—how a person is likely to react to disturbing images.

Destabilizing the Brain

I personally experienced the destabilizing effects of violent images several years ago when I lectured on the brain to a class of criminal profilers at the FBI Academy in Quantico, Virginia. At the conclusion of my lecture, several of the agents invited me to view and comment on a series of videotaped interrogations. Included among the videos was a serial killer and several convicted murderers who had committed horrifying crimes.

In order to convey as much information as possible, the agents presented the interviews on a split screen. On one side of the screen were the interrogator and the killer; on the other side was videotape of the crime scene in all its grisly detail. Especially jarring was the contrast between the matter-of-fact demeanor, calm tone, and general lack of emotion displayed by the killer with the stark images of blood and dismemberment on the other half of the screen.

Later that night I had trouble sleeping. Try as I could, I couldn't get those horrifying images out of my mind. I remained awake until I finally calmed myself down by taking a tranquilizer. Keep in mind that I'm a fully trained doctor with decades of personal encounters with blood, gore, and human suffering. Yet those videotapes disturbed me sufficiently to interfere with my sleep for several days. Here is what I think happened.

Our brain's right hemisphere is specialized for working with images, not words. For instance, if you're driving in unfamiliar surroundings while glancing down at a map, it's primarily your right hemisphere that processes the lines and figures on the map. But if, instead, someone in the passenger seat is telling you directions, your left hemisphere is the primary processor of the verbal description of your route. Of course, the hemispheres don't operate in complete isolation from each other: The streets and addresses on the map must be read by the left hemisphere and then the map's arrows and lines must be coordinated by the right hemisphere.

When encountering images of emotionally neutral or positive scenes the hemispheres remain in balance. However, when the image is one of horror, carnage, suffering, injury, or death, the balance of activity shifts toward the right hemisphere and the brain is in danger of becoming overwhelmed and dysfunctional. That's why watching images of disasters has a much more powerful effect on our mental stability than reading about the same events.

For example, a particularly disturbing image processed by the right hemisphere can overpower the rational, language-based operations of the left hemisphere. My response to the FBI videos resulted, I believe, from the overpowering affect those horrific images had on the balance between my brain's right and left hemispheres. Even though I knew that the horrifying images were part of an investigation carried out more than a decade ago and that the killer in the videos had been executed several years earlier, my right hemisphere's response to the images overwhelmed my left hemisphere's reasonable assurance that I had nothing to be upset about.

Several factors helped lessen the intensity and duration of my reaction to the episode at the FBI Academy. For one thing, I told the agents that while I would be happy to try to be of help by watching the interviews with them, I would prefer in the future not to watch the accompanying crime-scene videos. Second, I found relief from my distress by immersing myself in the technical details of the investigation and the fine points of profiling the perpetrator. Psychiatrists use the term "intellectualization" to describe this approach of displacing highly emotional material from awareness by concentrating on facts and techniques. This "defense mechanism" acts as a safeguard against being emotionally overwhelmed. And while intellectualization worked for me, I have often asked myself, "Suppose I had been forced to continue to watch these crime-scene videos? What would have happened then?"

You Can't Turn Off the Brain

Another reason images exert such power over our brain is that they are unmediated: A picture of a bleeding survivor of a terrorist attack is only slightly different from actually seeing that person—especially with the high definition images produced by digital cameras. A verbal description, in contrast, takes longer to read, understand, and finally—through an act of imagination—convert into an internal image, and since most of us have little personal experience with such things, that internal image is generalized, less focused, less riveting, less disorienting, less disruptive.

While words on a page open us up to ideas and communication with others, images do not reveal ideas as thoroughly as words do and offer only elemental communication with others. Many images seem only exclamatory in their value: "Look! How horrible!" Moreover, images exert a hypnotizing, gripping, seemingly irresistible influence on our behavior that words alone cannot sustain. For instance, in the first days after the terrorist attacks on New York City and Washington, D.C., millions of people sat before their televisions for hours, watching endless repetitions of the footage of the airplanes hitting the World Trade Center and the subsequent collapse of the towers.

"You keep watching because you think you might miss something. But what?" asked *Washington Post* staff writer Paul Farhi. "Something terrible always seems to be about to happen, piled on top of the terrible things that have already happened. So you hunt, orbiting the channels, long into the night, simultaneously fearful and eager. What you can't turn off is your brain." And not being able to turn off your brain exerts a cost.

When polled a week after the terrorist attacks, nearly three-quarters of Americans reported feelings of depression, half experienced difficulty concentrating, and one-third suffered from disturbed sleep. In addition, nearly two-thirds reported that they were "addicted" to watching replays and news about the attacks.

"Americans are saddened, frightened, and fatigued by what they are watching," concluded the report by the Pew Research Center for the People and the Press. *Washington Post* media critic Howard Kurtz was even more blunt: "The round-the-clock coverage of the attacks on New York and Washington, with its pictures of death and destruction and constant talk of a grueling war to come, is exacting a heavy psychological toll."

According to psychologist Roxanne Cohen Silver, who conducted research on the aftermath of September 11, watching the events live on TV produced levels of distress that in some cases equaled that of people who were either present at the sites or were in telephone contact with someone in the buildings or on the planes.

But even among people who don't ordinarily watch that much TV, repetitive exposure to televised scenes of horror can be destabilizing. One Washington, D.C., psychiatrist turned off her television in the aftermath of the terrorist attack because she felt herself getting too upset by the sight of people jumping from the World Trade Center. She then contacted her patients and suggested they limit their television viewing of the coverage.

"The visual medium is more intense," says Erica Wise, a clinical psychologist and associate professor at the University of North Carolina at Chapel Hill. "With radio, people do not have exposure to images that they will see again and again in their heads."

And images of violence can also elicit uncharacteristic behavior. Take Andaleeb Takafka, age 20, described by relatives and neighbors to a *Washington Post* reporter as "an unassuming young woman from an isolated town south of Bethlehem who sewed clothes in a factory and took little interest in politics." She detonated a suicide bomb on April 12, 2002, at a bus stop in Jerusalem three weeks into one of the Israeli incursions into Palestinian territory. The mother of this suicide bomber told the *Post* reporter, "Andaleeb watched television all the time. She watched all the killings, the tanks, the helicopters shooting. She would watch until two in the morning. I would tell her to go to bed. She would answer harshly, which was not like her."

Within two weeks of Andaleeb Takafka's suicide bombing, Milad Mohammed Hemeida, a 23-year-old Egyptian, walked towards Israeli soldiers announcing, "If anyone comes near me, I will blow myself up." After shooting him, the soldiers discovered that Hemeida was not carrying a bomb. In the weeks prior to his death, Hemeida spent much of his time watching cable-channel news footage of the Palestinian-Israeli conflict. Indeed, according to *New York Times* reporter Tim Golden, who investigated the shooting, Hemeida "was unusually consumed by what he saw on television."

Televised images of the aftermath of a suicide bombing can exert a similar effect on any of us. We are simultaneously horrified and *terrified* (the word terrorism comes from the Latin *terrere*, meaning to terrify). In his book *Terror in the Mind of God*, sociologist Mark Juergensmeyer describes his sentiments upon seeing pictures of a bus destroyed by a suicide bomber near Hebrew University—the same bus Juergensmeyer had taken the day before.

"The pictures of the mangled bodies on the Jerusalem street . . . therefore had a direct and immediate impact on my view of the world. What I realized then is the same thing that all of us perceive on some level when we view pictures of terrorist events: on a different day, at a different time, perhaps in a different bus, one of the bodies torn to shreds by any of these terrorist acts could have been ours."

Indeed, the images can be impressed so vividly upon the brain that they recur unbidden to wreak a psychic toll ranging from anxiety to post-traumatic stress disorder (PTSD).

A Psychic Toll

Included among the symptoms of PTSD are recurrent and intrusive recollections of an event that involved actual or threatened death or serious injury, coupled with distressing dreams of the event. But one symptom of PTSD stands out: the extreme response that occurs when the PTSD sufferer is reexposed to a situation originally associated with the traumatic event. For instance, a person who survived a serious car accident may reexperience the trauma whenever he or she drives by the scene of the accident.

This happens because the brain of a person afflicted with PTSD has lost the ability to suppress frightful and disturbing images related to the trauma. As a consequence, these images return in the form of flashbacks. The traumatized person literally relives the experience despite his or her best efforts to forget it. Fortunately only about 25 percent of people directly exposed to trauma develop PTSD. And while there is no conclusive evidence that watching traumatic situations on television can result in the disorder, preliminary research suggests that it does have an effect.

In August 2002, William E. Schlenger and his colleagues at the Research Triangle Institute in North Carolina published the first post–September 11 study of psychological symptoms and post-traumatic stress in New York, Washington, and the nation as a whole. They found a direct correlation between reports of post-traumatic distress and the number of hours spent watching the attacks and their aftermath. The more television watched, the more likely the person suffered psychological effects. And this correlation was unrelated to whether the affected person lost a friend or family member in the attacks.

Television viewing also turned out to be the critical variable among people directly affected by the attacks. According to a study of Manhattan residents carried out by Sandro Galea and colleagues at the New York Academy of Medicine, those directly affected by the attacks who repetitively watched televised images of people falling or jumping from the towers were six times more likely to develop PTSD compared to similarly affected people who did not watch TV (22.5 percent versus 3.6 percent).

Over the ensuing months after witnessing a disturbing event on television, the person afflicted with an acute stress reaction might have nightmares, anxiety, intrusive thoughts and images of the disaster, a sense of numbing or detachment, mood swings, irritability, and an over-reaction to trivial occurrences. The risk is particularly acute among people who live alone and spend a lot of their spare time watching television rather than socializing. For such people, TV provides a reality unrelieved by socializing with others, which can serve as a stabilizing force.

Emotional Shutdown

The brain, it turns out, can only process so much horror and mayhem before emotional shutdown occurs. Such a process was described by an Atlanta executive interviewed by a *New York Times* reporter in 1991 during crises in Bangladesh, Ethiopia, and Mozambique: "You can see real true-to-life pictures, but your mind reacts to it almost as if it's just a movie."

Or consider this response by Eastern European specialist John Fox after watching videos of a busload of refugee children shelled during the war in Bosnia: "The images just kept mounting. The images never stopped, and that's what got to people . . . you had to steel yourself just to get through the day."

I'm convinced that constant exposure to visual depictions of suffering, conflict, and violence creates dysfunctional circuits within areas of the brain that mediate emotion. The result? Various forms of PTSD; desensitization (MEGO, or My Eyes Glaze Over syndrome described by newspaper columnist Jack Payton); feelings of unreality and detachment that cause one to respond to real tragedy as if it were a movie; and, finally, various avoidance reactions ranging from phobias to emotional burnout.

When our emotional responses are expended on images of people and events that are remote from our own lives, does that reduce our capacity for empathizing with those close to us who are in need? Probably not, as long as we are in a position to actually do something positive for them. But while opportunities to provide aid may be available in real life situations, images permit only a passive response. In fact, watching images of the aftermath of suicide bombings or other emotionally arousing material may disconnect us from our desire and ability to respond with care and action. Constant exposure to such images may cultivate passivity in the face of events that demand empathy and action. Also, recycling images of violence and despair is becoming common. One day we're watching images from a riot, a car bombing, or the crime scene of a high-profile murder case; six months later, we're watching its televised recreation on a police drama or "reality" TV show.

While we all know that the televised recreation isn't "real," our emotional responses may be the same. Separating fact from fiction becomes even harder when we watch a show like *COPS*, in which the chases and the violence involve actual situations presented in a format that is basically a combination of fantasy and reality. Whether we're watching a dramatic recreation or actual footage of real police officers apprehending real criminals, our brain's right hemisphere processes the images the same, and our brain's limbic system provides the emotional responses that would be appropriate if the events were "really happening."

Walter Mitty and the Frontal Lobes

In a famous James Thurber story, the main character, Walter Mitty, compensates for his boring unchallenging life by imagining dramatic scenarios in which he is a hero. In some, Mitty physically overpowers various villains through acts of controlled aggression. But fantasized aggression isn't limited to Walter Mitty's world. On occasion most of us imagine ourselves physically lashing out at some annoying or irritating person.

Neuroscientists have recently discovered that a PET scan taken while we're entertaining violent daydreams within our imagination will show a decrease in the activity of the orbito-frontal cortex. This discovery resulted from some intriguing experiments carried out by cognitive neuroscientist Jordan Grafman and his associates at the NINDS.

In one of Grafman's studies, volunteers visually imagined three responses to the following scenario. The subjects are in an elevator with their mother and two men. Suddenly the two men assault the mother. In the first variation, the subjects picture themselves doing nothing. In the second they imagine themselves attacking the men but losing the battle and being physically restrained by one of the attackers. In the third variation—the Walter Mitty version—the subjects imagine themselves attacking the two men and hitting them with all their strength "to seriously injure or kill them."

PET scan images were obtained while the subjects visualized the three scenarios. While visualizing the two situations that involved an aggressive response, the subjects' brains showed a significant decrease in blood flow in the orbitofrontal cortex compared to the scenario that lacked any aggression.

"Abnormal expressions of aggressive behavior in violent individuals are accompanied by a functional 'shut down' of the orbitofrontal cortex," according to Jordan Grafman. The orbitofrontal cortex—the part of the frontal lobe most directly connected with the amygdala and the limbic system—is thought to play a pivotal role in integrating motivational and emotional processes. Aggression is accompanied by decreased activity in this area.

Grafman's study shows that in normal people, merely thinking aggressive thoughts is enough to alter blood flow and activation patterns in the part of their brains known to control angry impulses. And according to Marina Nakic, also at NINDS, "These findings suggest that the human medial prefrontal cortex mediates between the sight of an aggressive act and the elicitation of behaviors that such stimuli might trigger."

This means that either watching violence or just imagining it decreases the modulating influence of the prefrontal lobes on the limbic system. Admittedly, there is no scientific way of surveying the incidence of violent thoughts, but researchers can now estimate that, at least in terms of brain activity, there is a relationship between watching violent images, thinking about acting violently, and subsequent violent behavior.

Within weeks of my discussion with Marina Nakic, Jeffrey G. Johnson of Columbia University released his study of the effects of television viewing on violent behavior. By examining the television viewing habits of 707 families in upstate New York,

Johnson found that adolescents and young adults who watched more than seven hours of television per week were more likely to commit an aggressive or violent act in their later years.

Johnson's study, the longest of its type, began in 1983 when the average age of the participants was 14. Eight years later, in 1991, Johnson correlated television-viewing statistics with law enforcement records of violence. Among those who watched one to three hours a day, 18.4 percent had acted violently (defined as resulting in a serious injury, such as a broken bone). Among those who logged in more than three hours a day, the rate of violence was 25.3 percent.

Another eight years later, in 1999, Johnson reinterviewed the subjects and once again tallied their incidence of violent behavior. While 1.2 percent of the adults who watched less than one hour per day had acted violently, 10.8 percent of those who watched three or more hours had assaulted someone with sufficient ferocity to create a bruise or scar.

(Since the propensity for violence usually decreases as males exit adolescence, an overall reduction in violence among both groups from the 1991 study would be expected.)

Notice that Johnson's study didn't confine itself to the time people spent watching *violent* television fare but simply the total time watching television. Two possible assumptions can be made. First, violence on television is so pervasive that it's reasonable to assume that frequent watchers are automatically exposed to it. Second, perhaps simply sitting in front of the set, no matter what may be playing at the time, leads to a later propensity for violence.

Personally, I find the second possibility rather far fetched. The first assumption seems much more likely: If a person watches enough television, he or she is inevitably exposed to violence, since well over half of the programs on television depict violent acts. Consequently, there is a strong correlation between the time spent watching television and the amount of violence viewed. And that data seems to indicate that watching violence predisposes some people to act violently.

"There may be a vicious spiral among at-risk youths in which media violence contributes to the development of aggressive attitudes and feelings, which promote further viewing of media violence, increasing the likelihood of aggressive behavior," according to Johnson.

Johnson and his colleagues speculate that even among youths not predisposed to violence, the televised depiction of violence may desensitize viewers and prompt them to assume that violent behavior is acceptable. The suggested solution is straightforward: "By decreasing exposure to media violence we may be able to prevent millions of Americans from being raped and murdered," Johnson told *New York Times* reporter Gina Kolata.

Johnson's research, combined with discoveries of the brain's response to real or imagined aggression, suggests that watching depictions of violence lowers the threshold for the expression of violence. And the correlation is a strong one.

"The correlation between violent media and aggression is larger than the correlation between exposure to lead and decreased I.Q. levels in kids," according to Brad Bushman, a professor of psychology at Iowa State University in Ames, who contributed a commentary accompanying Johnson's study in the journal *Science*. "It's larger than the effects of exposure to asbestos. It's larger than the effect of secondhand smoke on cancer."

Immature Brains May Not Discern Between Real Violence and Images of Violence

More than 1,000 other studies support Johnson's conclusion that when children watch violent media they become more aggressive. Conversely, aggression can be decreased by as much as 25 percent simply by decreasing a child's exposure to television and movie violence. This finding by pediatricians and psychiatrists, published in the *Archives of Pediatrics and Adolescent Medicine*, is partially explained by the immaturity of the child's brain. While most adults can distinguish media images from real images, a child has greater difficulty doing that. His or her right hemisphere processes the disturbing image but—due to left hemisphere immaturity and lack of life experiences—can't put the images into context.

"There are three effects on children of viewing violence on television," according to John Murray, a professor of developmental psychology at Kansas State University in Manhattan. "Children may become desensitized to the pain and suffering of others. They may become more fearful of the world around them, and they may be more likely to behave in aggressive or harmful ways towards others."

Typical of an aggressive response is eight-year-old Calla Perkins, a second-grade student at Public School 89, near Ground Zero of the World Trade Center attacks. Eight months after September 11, a *New York Times* reporter described Calla as "bursting with anger" and expressing "fury in ways she would later regret." She kicked her cats and dog, as well as a sign advertising World Trade Center pictures. "I'm so angry I don't know what to do," Calla told her mother.

"We should shield children from TV," according to Joanne Cantor, author of *Mommy, I'm Scared*. "The images on TV are potentially much more upsetting than in the newspaper because the way things are presented on TV is very visual and very emotional."

Young children, as I've said, cannot reliably distinguish between something happening for real and something happening on a video replay or dramatic depiction. To 5- to 10-year-olds, events occurring on the TV are always happening *now*. Young children are also easily confused about geographical limitations and may mistakenly assume things about where an event is occurring and how many people it is affecting.

For example, the eight-year-old child of a friend of mine living in the suburbs outside New York City reacted to the September 11 attack by refusing to separate from her mother in the morning to attend school. She begged her mother not to return to her job as a publicist for a major New York publisher. Uncertain what to do, my friend decided one morning to take the child to work with her. Later that day her daughter spontaneously remarked that prior to her trip to the city she had imagined that all of New York City had been reduced to the crater shown on television the previous week. Relieved that everything near her mother's office appeared unchanged, she stopped expressing anxiety about her mother going to work in the city.

Also at great risk when viewing violence on television are people—children and adults—who are already showing signs of emotional disturbance or who suffer from mental illness, especially anxiety disorders, which affect more than 19.9 million Americans. For somebody already experiencing low levels of chronic anxiety, watching something like the aftermath of a terrorist attack can lead to an acute outbreak of disabling symptoms.

Whether you're comfortable with the idea that media depictions of violence incite violent acts in the viewer, the neuroscientific research is pretty clear: Thanks to the work of Grafman, Nakic, and others, we know that watching violence—or even just imagining it—reduces the functional activity of those parts of our brain that are normally enlisted to inhibit violent impulses. And while this reduction may exert little influence on people for whom violence will always remain an unacceptable path for resolving conflict, the reduced prefrontal activation may exert a powerful influence on people with abnormal frontal lobes.

For example, people with antisocial personality, a disorder associated with increased aggressive and violent behavior, have an average 11 percent less gray matter in their prefrontal cortex compared to healthy people. When such a person looks at vivid depictions of violence, he (most antisocial personalities are men) is already compromised in terms of his ability to control his violent impulses. Add to that the deactivation found by Grafman and Nakic in their research and you have a potentially deadly combination of factors that can lead to a violent outburst.

Normalizing Our Emotions

The good news is that we retain a good deal of control over the images we watch. We can shut off most images that we find disturbing. Does this mean that we become insensitive and emotionally unavailable to the suffering of others? On the contrary. By limiting our exposure to images of horrible situations that we can do little about, we become more sensitive to the sufferings of those around us whom we can help. When we are not "burned out" or desensitized by media-derived emotions, we can reach out to others. Only by taking steps to normalize our emotional reactions can we be meaningfully available to others. Our goal is to achieve a balance between the rational and emotional networks in our brain.

Just as we don't want to base all of our responses on purely rational considerations and become like robots, we don't want to allow our emotions to become so ascendant that we become nonfunctional. We can help this process by controlling the imagery we allow into our brains.

One thing seems indisputable: Nobody has ever claimed that watching violent material leads to a *lessening* of violent impulses. And since most of us are turned off by vivid depictions of murder, assaults, battering, and child abuse when they occur in real life, why do we allow our brains to be assaulted by video recreations of these things?

Over the next decade we can expect further refinements in our understanding of the brain bases of violence, along with a greater ability to predict violence in those predisposed to it.

From a practical point of view, it makes a lot of sense for you to avoid vivid images of events that, according to what we're learning from new brain research, can lead to psychological harm. By eliminating media depicted violence and taking control over the images you incorporate into your brain, you can increase your well-being and happiness even under conditions of stress and loss.

chapter 5

The Happy Brain: The Joy and Music in You

In the previous chapters we explored the brain changes associated with some of life's more troublesome experiences, such as distraction, fear, depression, anxiety, anger, stress, and violence. But the New Brain is also the mediator of positive emotions and experiences, such as humor and laughter.

Scientists have long known that humor exerts a positive effect on our general functioning, perhaps even helping us stave off the ravages of certain diseases.

The late Norman Cousins described in his best-selling book, *Anatomy of an Illness as Perceived by a Patient*, how he used humor to recover from a mysterious and rapidly progressive disease. With the approval of his doctor, Cousins checked into a hotel instead of a hospital and for several hours each day watched old Marx Brothers films and reruns of *Candid Camera*.

"I made the joyous discovery that 10 minutes of genuine belly laughter would have an anesthetic effect. Laughter not only provides internal exercise for a person flat on his or her back but it creates a mood in which other positive emotions can be put to work. In other words, laughter makes it possible for good things to happen."

In the ensuing 25 years since Cousins's book, scientists have confirmed that humor decreases stress, boosts immune defenses, relaxes muscle tension, decreases blood pressure, and modulates pain. But what specifically happens in your brain when you respond to a joke with laughter?

First, when you listen to the joke you have to pay attention and follow the story to where the humor becomes apparent. In order to "get" the joke you must trace the abstract and literal meaning of the words and then shift your perspective to appreciate the incongruity that lies at the basis of most jokes. Your working memory compares the narrative with the punch line at the end. To do this you must hold the details of the joke "online," that is, within your working memory. But more is required than just mentally appreciating incongruities. In fact, humorless people often get the point of a joke—they just don't find it funny.

A good joke also engages the body. Smiling and laughing activate the muscles of the face and throat and help produce the sound of laughter. That sound varies in loudness and intensity from one person to another and from one joke to another. Indeed, a smile unaccompanied by the sound of laughter usually implies that for one reason or another the intended audience isn't amused. Certainly a standup comedian won't remain on the stage for long if his audience sits in motionless silence. But when the comedian tells a sidesplitting joke, the audience comes alive with laughter, noise, and movement.

The Pleasure Pathway

To understand what happens in the brain when we laugh, neuroscientists have used imaging techniques to study people reading jokes, looking at cartoons, and listening to recorded laughter. They have discovered what appears to be a humor-processing circuit that involves parts of the frontal lobe, especially the supplementary motor area (SMA) and the nucleus accumbens, a component of the "pleasure pathway."

The SMA, located along the inner area of the frontal lobes, processes both the cognitive aspect of humor (getting the joke) as well as the motor part, such as facial movement and breathing (smiling and laughing). Fibers from the SMA transfer a program for laughter to the overmotor and motor areas.

In addition to enhancing motor activity, jokes simply make us feel good. We like to hear them, and some of us are fond of telling them. The instant we hear or tell a joke we experience an inner sense of well-being. Things don't seem so serious or so humor*less* anymore. This feeling comes from the nucleus accumbens, an important way station within the "emotional brain" that consists of a circuit of linked structures. The circuitry for humor and laughter also includes the a part of the hypothalamus called the anterior cingulate cortex (ACC), the temporal lobe and its connections to the amygdala, and the junction within the brainstem of two coordinating structures called the pons and the medulla.

Laughter can even be induced by electrical stimulation of the deep brain center, the subthalamic nucleus. The stimulation elicits mirthful laughter and hilarity in patients undergoing stimulation at that site as a means of improving their Parkinson's disease. Patients with tumors and disturbances elsewhere in the brain sometimes break into uncontrollable laughter under circumstances that would not ordinarily elicit such a response. In one famous case, a young boy began laughing uncontrollably while attending a funeral. His "pathological laughter"—the neurological term for the disorder—had resulted from a burst blood vessel within the brain's third ventricle.

Automatic laughter can also be evoked from the cortical areas closer to the brain surface. A 1998 report in the journal *Nature* describes a young epileptic woman who smiled and then laughed while undergoing electrical stimulation of the SMA.

People with frontal lobe damage (especially in the right frontal lobe) have the opposite reaction—they rarely laugh spontaneously, fail to appreciate humor, and laugh and smile less when hearing a joke. They also don't react humorously to cartoons, since frontal lobe damage interferes with the ability to focus attention and visually explore the cartoon for the important details that render it humorous.

A similar absence of humor occurs in people with Alzheimer's disease. This results from damage to areas of the brain important in the formation and maintenance of memory. Damage in the hippocampus, which is common in Alzheimer's victims, interferes with encoding the details of a joke, while frontal lobe damage impairs the working memory required to link the material from the beginning of the joke to the humor at the end.

"Laughter has many subtle effects on our social companions," writes behavioral biologist Silvia H. Cardoso in the journal *Cerebrum*. "It breaks the ice, achieves closeness, bonds us, generates goodwill, and dampens hostility and aggression. Observe how we laugh when we want to deflate tension between strangers or need to say no to someone. Laughter disarms people, creates a bridge between them, and facilitates amicable behavior."

Tickling is an example of the social nature of laughter. Try tickling yourself. It doesn't work, does it? Charles Darwin concluded that tickling elicits laughter only when done by someone else and when the precise point of stimulation cannot be known ahead of time. Our brain is so good at distinguishing between self-initiated sensations and sensations initiated by others that we even laugh when a robot carries out the tickling, but we won't laugh if we control the robot's tickling movements.

Here are two questions about laughter that you may be curious about. Is the brain organized any differently among those with a good sense of humor? And if we make deliberate efforts to cultivate a sense of humor can we improve brain function? Although I suspect the answers to both questions are "yes," there isn't any proof for it at the moment. But expect to hear a lot more over the next decade about the brain circuits involved in humor and the beneficial effects of laughter.

Music and the Brain

Should music be considered merely "auditory cheesecake," as one commentator put it, a nice but unnecessary addition to our daily lives, or does it serve a more important purpose?

Whatever our response, one thing is certain: Music exerts a powerful and measurable effect on the brain. In some instances music can even trigger "musicogenic seizures," which are epileptic attacks that occur while listening to music that a person finds arousing or inspirational.

For some of us, life would be unbearable without music; for others, music doesn't seem to make that much difference one way or the other. For a person afflicted with *congenital amusia* (nearly complete tone deafness), music isn't distinguishable from other sounds. About 5 percent of the population suffers from this syndrome. Victims usually have normal hearing, intelligence, and general neurological functioning. The revolutionary leader Che Guevara suffered from amusia. When attending dance parties he had to rely on companions to help him distinguish between a tango, with its slow gliding movements and abrupt pauses, and a gentle Brazilian samba.

I learned of another example of amusia from Catherine Reed, a psychologist at the University of Denver. Her patient, M.G., is an educated 63-year-old man who, during childhood, received extensive musical training. Despite his familiarity with music, he describes it as "structured noise" or "the sound of a car door slamming." Imaging of M.G.'s brain by fMRI reveals impairment for complex aspects of music such as appreciating multiple tones, harmony, tone sequences, and melodies. As a result, M.G. cannot recognize nor sing songs that he's been exposed to since childhood. Even when listening to old stalwarts like "Happy Birthday," he never hears anything amiss, even if the song is so off-pitch most people would wince in discomfort. Some amusics (M.G. is an exception here) also experience difficulty in detecting prosody (intonation and stress) in normal speech. Thus, they often don't "get" sarcastic comments or other subtle meanings expressed by tones of voice or word emphasis such as "He's a *real* genius" when "real dullard" is the intent conveyed by the speaker's tone of voice.

Aside from such rare conditions as amusia, music provides most people with intellectual stimulation and enhances powerful emotions—ranging from feelings of "shivers down the spine" to unity with the cosmos. While we are listening to or playing some kinds of music, our heart rate slows and muscle movements diminish to the point of almost total immobility. But different kinds of music invoke strikingly different responses in the brain. Music played in a major key and with a quick tempo brings about physical changes typically associated with happiness, such as a slight increase in breathing rate. A slow tempo played in a minor key elicits physical changes associated with sadness, such as a slowing of the pulse, a slight rise of blood pressure, and a mild temperature elevation. These responses result from activation of a brain network that includes the amygdala, parts of the pre-frontal cortex, and other areas involved in reward, emotion, and motivation.

A Uniquely Human Invention

"The weight of the evidence is that music is a uniquely human invention . . . among the most complex of human activities, involving perception, memory, timing, grouping, attention, and (in the case of performance) expertise and complex coordination of motor action," according to Daniel J. Levitan, a psychologist with the department of music theory at McGill University in Montreal.

Levitan suggests that music exercises our brain by challenging it to discern different patterns and groupings. "Patterns emerge, regroup, repeat, and fold in on themselves in many interesting ways," he says. He gives as an example Beethoven's "Moonlight Sonata," which "moves us because each time we listen to it, we hear it slightly differently, depending on the performer, our mood, the people who are with us. The notes of the composition form a foreground, the spaces between them a background, and our brains work actively to make links between them, group the music into phrases, and predict what will come next. The ever-vigilant brain becomes excited when we encounter even subtle violations of our expectations."

Levitan and others are employing the tools of neuroscience to discover what is happening in the brains of musically sophisticated individuals and how it differs from those who know or care little about music.

In general, musically sophisticated individuals are more likely to perceive music in an analytical manner and thus rely more heavily on their dominant (usually left) hemisphere. That is, if you have learned to play an instrument or studied music to the point that you can meaningfully compare different performances, you are more likely to engage in analysis and comparison—left hemisphere functions. But if you've received little formal musical training, you don't rely much on analysis. You simply prefer some types of music to others and don't consciously think much about the reasons why—a predominately right hemisphere response. This division, incidentally, isn't a rigid one. Hard and fast rules aren't possible, since great variability exists between one person and another in terms of musical interests, abilities, and skills. Simply learning to play a musical instrument may shift the balance from right to left hemisphere functioning.

As an example of how musical training influences brain circuitry and enhances emotions, consider the auditory cortex, which is located slightly above the ears. Operationally, the auditory cortex occupies a privileged place midway between the frontal lobes—just behind the forehead—and the deeper dwelling limbic system components. Thanks to the two-way connections between the auditory cortex and the frontal lobe, we can intellectually appreciate, compare, and judge a musical performance. And thanks to the two-way connections between the auditory cortex and the limbic system, we can respond to the emotional resonance that accompanies all great musical compositions. Not surprisingly, the auditory cortex is larger in musicians compared with the brains of non-musicians.

Other brain areas that are larger in adult musicians are the primary motor cortex and the cerebellum (both of which are important in movement as well as coordination); the corpus callosum (the thick fiber track uniting the right and left hemispheres of the brain); and an area at the surface of the temporal lobe called the planum temporale.

Enlargement of the left planum temporale occurs in musicians as well as those with perfect pitch, according to the findings of Takashi Ohnishi of the National Center of Neurology and Psychiatry in Tokyo. Ohnishi believes an enlarged planum temporale allows musicians to hear music differently from other people. Instead of merely hearing music they are analyzing it.

With the exception of the planum temporale, musicality resides primarily on the right side of the brain. We know this from tests performed on patients who have had one or the other temporal lobe surgically removed. Right temporal removal resulted in difficulties in recognizing changes in key (the set of pitches corresponding to the notes in a melody) and contour (the rising and falling pattern of notes making up a melody).

Researchers have also uncovered functional changes. In one study from the University of Tübingen in Germany, professional and amateur musicians underwent fMRI scanning while they executed the fingering used to perform the first 16 bars of Mozart's Violin Concerto in G Major. This selection was chosen because it is often performed by professional violinists and yet is easy enough for many amateurs to play.

"We found that activity in the part of the [primary] motor cortex that controls finger movements was much more economic and focused in the brains of the professional musicians," according to Gabriela Scheler, the lead investigator in the study. "This decrease in overall activity suggests that the professional musicians are using their brains more efficiently, presumably because they've practiced the concerto so many times that playing it has become automatic."

In the auditory and prefrontal areas of the brain, however, the professionals showed *increased* activity compared to amateurs. "When the professionals move their fingers they are also hearing music in their heads, so that's why the auditory region is more active. Presumably, this enhances musical performance," concludes Scheler. The increased prefrontal activation found in the professionals is thought to correlate with working memory responsible for encoding and integrating the finger movements established during previous performances.

Interestingly, this pattern in professional musicians of decreased motor activity and increased auditory and prefrontal activity also occurs if the musicians are asked to imagine playing the concerto without moving their fingers.

Years of Practice and Performance

Are the brains of skilled musicians different as a result of their years of intense training, or did those differences exist prior to their taking up music? In other words, are the brains of people destined for musical accomplishment "wired" differently from the start?

"If musical training affects the development of the brain, we would expect that the brain responses evoked by musical stimuli in highly skilled musicians would differ from the brain responses of non-musicians," says Larry Roberts of the department of psychology at McMaster University in Hamilton, Ontario.

To check this hypothesis Roberts and his colleagues carried out a two-part experiment. In the first part they recorded the EEGs (brain wave activity) of highly skilled violinists who were members of Canada's National Academy Orchestra and compared their brain wave recordings to recordings from a group of non-musicians. During the recordings both groups listened to violin tones, piano tones, and nonmusical pure tones of the same pitch as the musical tones. Roberts found that the musical tones activated more neurons in the auditory cortexes of the musicians than in the non-musicians.

"But our study comparing the musicians and non-musicians still didn't tell us whether the brain attributes we found in the musicians are caused by the musical training they received," reasoned Roberts.

To answer this question, Roberts trained the non-musicians to detect very small changes in pitch. After only three weeks of training their brains also showed an increase in activation in the auditory cortex. This suggests that the brain changes observed in highly skilled musicians aren't inherited but result from many years of practice and performance.

As Roberts's research illustrates, brain modifications can sometimes take unusual directions or develop in unexpected ways: Non-musicians required only a modicum of training for their brains to begin to respond in ways characteristic of professional musicians.

It should be possible in the near future to design imaging studies to assess musical abilities. For example, an initial study done in childhood could serve as a standard for later comparison when the musician has reached his or her musical maturity. When available, such technological measurements will herald new unities between art and brain science. Rather than providing an explanation for musical talent, however, the imaging studies will serve as a basis for measuring it. But whatever technological advances may occur in the future, I'm confident that neuroscience will never be capable of providing a total explanation for our appreciation and enjoyment of Mozart's Piano Concerto no. 9.

Nor will neuroscience demystify the process of musical creativity. Although there are several reasons for this, one in particular stands out. Music and the brain involve different orders of discourse. While it's true that we wouldn't be able to appreciate and perform music without contributions from specialized areas within the brain, we cannot hope to understand the seven different attributes or "dimensions" of music (pitch, rhythm, tempo, contour, timbre, loudness, and spatial location) in terms of electrical currents across nerve cell membranes or the transport of neurotransmitters across synapses.

Instead of neurological *explanations* of music, we will be able to draw important *correlations*. Think how exciting and intriguing it will be when neuroscience shows us in fine detail some of the processes taking place in our own brains as we play a musical instrument or listen to a master employ that same instrument at a level of performance that we can only dream of achieving.

chapter 6

Modern Imaging Techniques: Windows on the Mind

Will brain research someday provide reliable measures of deeply hidden, unconscious attitudes that conflict with a person's statements about himself or herself? Will neuroscientists develop a reliable technique for detecting when someone is telling a lie? These are two of the practical questions neuroscientists are approaching, made possible largely through advances in technology.

Attempts to develop lie detection technology can be traced back to the 1920s, with the invention of the polygraph. The polygraph measures physiological indications of lying, such as alterations in heart rate, breathing, and galvanic skin response, or changes in electrical resistance of the skin. Although sometimes helpful, polygraphs have two main weaknesses. First, they measure changes taking place downstream from the brain, in the peripheral nervous system, rather than the brain itself. Second, some people can control their autonomic responses and thereby frequently pass the test even when telling a lie. For these reasons neuroscientists have long sought direct measurements of brain function associated with lying.

One method devised by psychiatrist Daniel Langleben of the University of Pennsylvania in Philadelphia, uses functional magnetic resonance imaging (fMRI). Like all lie detection methods, it's based on what psychologists refer to as "the guilty knowledge test." In essence, a person can only lie when he knows the correct answer but chooses to pretend that he doesn't. "You have to know the truth in order to deceive," according to Langleben. As Saint Augustine put it centuries earlier, "Deception is denial of truth."

In Langleben's experiments, his subjects are made aware of certain information (a card that they have looked at briefly and then pocketed, for example). Then, while lying inside an fMRI machine, the subjects peruse a series of cards including a duplicate of the one they have pocketed. When asked, the subjects deny previously seeing any of the cards and attempt to control their voice and demeanor so that nothing gives them away. But the fMRI detects the lie. When the subject sees the card identical to the pocketed one, his fMRI depicts increased blood flow (a measure of increased activity) in the anterior cingulate cortex, or ACC.

"The activation of the anterior cingulate with lying makes sense since this brain area is known to play a prominent role in error monitoring and inhibiting responses," according to Langleben.

When lying, the subject's ACC helps to inhibit the natural truth-telling response. This inhibition requires energy, hence the diversion of blood and oxygen to that part of the brain.

Brain Fingerprinting

While the anterior cingulate cortex research relies on displaying blood flow and oxygen usage within the brain depicted in the colored patterns in the fMRI, another method, informally known as Brain Fingerprinting, works by detecting changes in the brain's electrical activity, which varies depending on whether something is recognized as familiar. For instance, if you are shown a picture of the front of your house or apartment building, your brain's instant recognition of this familiar site will trigger an electrical response that will differ from your brain's response to an image that is unfamiliar.

In 1993 the inventor of Brain Fingerprinting, Lawrence A. Farwell, tested his technique at the FBI Academy in Quantico, Virginia. His goal was to distinguish 11 FBI agents from four imposters by measuring brain responses to cues that would be familiar only to a trained agent.

Each subject in the experiment wore a headband equipped with special sensors. While seated in front of a computer, the subject watched a series of words and pictures flash briefly on the screen. Some of the words and pictures were "targets"—things that the average person would be expected to recognize, such as the first few lines of the Lord's Prayer or a picture of the American flag. "Probes," on the other hand, were specific acronyms or terms that would be recognized only by an FBI agent. The third class of stimuli—"irrelevants"—were unlikely to be recognized by any of the subjects. To make the test even more challenging, the FBI agents were instructed to do everything they could to conceal from Farwell that they were in any way connected with the bureau. In each case, by monitoring the electrical activity in each subject's brain Farwell was able to correctly distinguish between the real agents and the imposters.

The principle behind Brain Fingerprinting is surprisingly simple. The memory centers of the brain respond to the sight of something familiar with a distinct change in electrical activity. This is not a new observation, incidentally. Neuroscientists have known for more than 30 years of the existence of these extremely brief electrical wave patterns. Farwell refers to all of these changes in electrical activity as a MERMER, an acronym for the jaw jarring "memory and encoding-related multifaceted electroencephalographic response."

Brain Fingerprinting may find its biggest application in criminal investigations. When a person commits a crime, his or her brain records details of the crime and stores them in memory. This process doesn't require conscious intent or willingness—indeed the criminal will try his best not to remember the crime lest his knowledge of the details trip him up during interrogation. But regardless of the criminal's denials, his brain made at the time of the crime the equivalent of an audio and video recording and stored it in memory. What's more, that memory persists for years.

Take the case of James B. Grinder, a woodcutter suspected in

1984 of murdering Julie Helton, a 25-year-old Missouri employee of a book publishing company. On August 5, 1998, Farwell tested Grinder's brain for details about the 14-year-old murder. Six days after crime-scene details were shown to Grinder and Brain Fingerprinting tests revealed his knowledge of the crime, Grinder pleaded guilty to first-degree murder. He was sentenced to life in prison without the possibility of parole.

"His brain said that he was guilty of the murder of Julie Helton," says Farwell. "He had critical, detailed information only the killer would have. The murder was stored in his brain and had been stored there 15 years ago when he committed the murder."

As with polygraph evidence, "brain prints" aren't accepted as evidence in court. Some of the objections to them involve disagreements about how to prepare and present the questions and images. For instance, in some highly publicized crimes, television and newspaper coverage could result in an innocent person knowing quite a few specific details of a crime. More important, additional research is required to establish with certainty that a guilty person can't outsmart the technology. Despite the uncertain future Brain Fingerprinting has in determining guilt in criminal matters, brain monitoring technologies aren't limited to evaluating truthfulness.

Reading Minds

Most of us at least occasionally act in illogical or irrational ways. Although we know what we should do, we do just the opposite.

Can brain research provide any help here? Yes, according to psychologists William Gehring and Adrian Willoughby at the University of Michigan in Ann Arbor. In their experimental study published in the journal *Science*, volunteers donned electrode caps that recorded event-related brain potentials (ERPs). ERPs, like MERMERs, are changes in brain electrical activity in response to a specific stimulus.

Gehring and Willoughby measured ERPs while their subjects engaged in a betting game played on the screen of a computer. Two boxes were displayed—one box represented a 5-cent bet; the other a 25-cent bet. After the subjects made a selection, the boxes changed color. If the box selected by the gambler turned green, the amount in that box was added to the player's winnings; if it turned red, the amount was subtracted from the winnings. A few seconds later, after the screen reconfigured the boxes, the subjects chose again.

Win or lose, the subjects' brain waves displayed greater activity (signaled by a downward dip in the brain wave recording) from the medial frontal cortex. But this medial frontal negativity (MFN) showed the greatest dip after loses—a response that occurred within 200 to 300 milliseconds (about a quarter of a second) after the gambler learned the outcome of each bet. During a string of losses, the MFN dipped further with each subsequent loss.

Gehring and Willoughby suggest this increased MFN activity corresponds to the well-known gambler's fallacy: that consecutive losses mean a gambler is "overdue" for a win, which fuels escalations in the amount of money bet. In other words, there is a

correlation between the gambler's tendency to bet more following a string of losses and the degree of dip that occurs in his
or her brain's MFN.

"Losses loom larger than gains," write Gehring and
Willoughby. "The aversion to a loss of a certain magnitude is
greater than the attraction to a gain of the same magnitude. After
a loss the brain thinks it's due for a win. As a result, when we
make a quick decision and it turns out to be wrong, we tend to
take a bigger risk the next time than we would have if our first
choice was right, and we can evaluate the outcomes of these decisions before we've consciously thought about what we're
doing."

Freud wrote about unconscious mental processing that sometimes lead to decisions made without recourse to conscious deliberation. But Freud tethered his observations to controversial
and debatable theories about the importance of underlying aggression and sexuality. Now neuroscience is revealing, instead,
that unconscious mental decisions, like the gambler's fallacy, result from a rapid quarter-of-a-second computation. Further,
these split-second, often impulsive decisions about how well
things are going are linked to a distinct pattern of activity (MFN)
carried out by the ACC.

The MFN study represents an early step toward understanding human subjectivity within a neuroscientific basis. Further developments will depend on designing experiments that
combine behavioral observations with recordings of the electrical
events within the brain. As with the Gehring and Willoughby
study, we may find that the electrical events take place in frac-

tions of a second—far too quickly for anything approaching conscious deliberation.

For example, imagine yourself rapidly scanning a series of pictures of human faces. As you look at some of the pictures, a loud burst of unpleasant noise plays in the background. Not surprising, your brain responds to the noise with mild fear. But after a while your brain becomes conditioned and automatically responds with fear when certain faces are shown, those paired with the unpleasant noise.

But suppose the pictures are shown so rapidly that you don't consciously perceive them. Shouldn't that eliminate the mild fear response? Actually, while some parts of your brain will respond less or not at all, the response from the amygdala will be the same whether or not you are consciously aware of the pictures. It's as if the conditioning process sets the amygdala at a higher sensitivity.

A similar response occurs in survivors of horrific events such as the September 11 attacks. "When you scan them, you discover that their amygdalas are switched on to far higher levels than you would find in the average person," according to Rashid Shaikh of the New York Academy of Sciences.

In short, functional neuroimaging is providing us with new ways of looking at how our emotions are programmed within our brains. Thanks to fMRI and other techniques, the nerve cell networks responsible for our emotions are being mapped out, most notably within the circuitry of the amygdala and anterior cingulate cortex. This imaging of thought and emotion has practical implications for how we reach everyday decisions.

"Emotional and rational parts of the brain may be more

closely intertwined than previously thought," according to Dean Shibata, assistant professor of radiology at the University of Washington in Seattle. In his research, Shibata scans the brains of subjects while they are in the process of making decisions. He finds that it makes a good deal of difference in the brain whether the decisions involve choices that affect health or well-being—decisions such as, Should I put on a seat belt? Should I make an appointment for a physical examination?

"Our imaging research supports the idea that every time you have to make choices in your personal life, you need to 'feel' the projected emotional outcome of each choice—subconsciously or intuitively," says Shibata. "That feeling guides you and gives you a motivation to make the best choice, often in a split second."

Shibata finds that making decisions that affect you personally enhances activity in part of your frontal lobes (the ventromedial frontal lobe). As a rule, you do not activate that area when thinking about events that do not involve you personally. "When people make decisions that affect their own lives, they will utilize emotional parts of the brain, even though the task itself may not seem emotional," says Shibata.

Know Thyself

Philosophers have spoken of self-knowledge as our most important goal in life: "Know thyself," as Socrates put it. Faced with making important decisions, we want to remain "reasonable" and "rational." We say we don't want our emotions interfering with our judgment. But keeping our reasoning powers uncontaminated by our emotions isn't as easily accomplished as we have been led to believe. Many times the influence of our emotions on our reasoning impedes self-knowledge—in moral dilemmas, for instance.

Imagine yourself faced with the following dilemma. A runaway trolley is bearing down on five people who will be killed unless you find a way to change the trolley's course. You can choose to turn a switch that shuttles the trolley onto another track where a worker is carrying out some needed repairs. Although the trolley will kill that man, the five other people will be saved. What would you do? Most people would reluctantly turn the switch in order to save five lives at the cost of losing one life.

Now consider a variation to the dilemma. You are standing on a footbridge overlooking the tracks at a point between the trolley and the five people. Beside you is a stranger—a portly, perhaps slightly inebriated stranger—who could serve as a convenient means of stopping or at least slowing the trolley if you pushed him from the footbridge onto the tracks below. Would you be willing to deliver that final fateful shove to save the lives of the other five people?

Despite the dramatically different scenarios, the math is the same for both: one person killed in order to save five. But though the numbers are the same, most people could not deliberately deliver the death-dealing push to the stranger, even though many would be willing, albeit with some reluctance, to switch the trolley onto the alternate track dooming the repairman. What's going on here?

Philosophy students have wrestled for centuries with such moral dilemmas. Reason isn't much help here since, as I mentioned, the math is the same. Thus, the philosophers face a dilemma: If reason doesn't provide a solution, what is the best approach to resolving this dilemma?

It may help to shift our attention to the emotional component. Most of us simply can't imagine ourselves, under any circumstance, pushing someone to a certain death in the path of a speeding trolley. Why? Because deep down inside most of us *feel* that such an act is fundamentally and morally different from simply flipping a switch. In short, our preference for pushing a switch rather than pushing a person rests on emotional, rather than strictly rational, factors. But how do we prove the truth of this intuition?

In an experiment conducted by Joshua Greene of the Center for the Study of Brain, Mind, and Behavior at Princeton University, subjects imagined the "push the person" scenario while undergoing an fMRI. Their fMRI patterns showed activation of brain areas associated with emotion. These areas—the medial frontal gyrus, posterior cingulate gyrus, and angular gyrus—are known to become active during states of sadness, fright, or general uneasiness. These areas, however, didn't light up with the "push the switch" scenario.

"The crucial difference between the trolley dilemma and the footbridge dilemma lies in the latter's tendency to engage emotions in a way that the former does not," according to the researchers who reported their study in *Science*. "The thought of pushing someone to his death is, we propose, more emotionally salient than the thought of hitting a switch that will cause a trolley to produce similar consequences, and it is this emotional response that accounts for people's tendency to treat these cases differently . . . and these differences in emotional engagement affects people's judgments."

Advances in neuroscience will allow us to conduct similar experiments that will clarify the role of emotions in some of the ethical and moral dilemmas we face in our lives. Indeed, dilemmas may not be required. Conflict alone may turn out to be a pivotal variable.

Dynamic Vigilance

With imaging technology, neuroscientists have recently pinpointed the part of the brain activated by conflict—the anterior cingulate cortex—which is located toward the front and middle of the cerebral hemispheres and directly above the corpus callosum. The ACC is, according to an early student of the region, "the seat of dynamic vigilance by which environmental experiences are endowed with an emotional consciousness." This dynamic vigilance is lost after damage to the ACC.

A person with injury to the anterior cingulate cortex often stops talking, ceases all spontaneous movement, shows little emotion, and appears lethargic and apathetic. As healing progresses, these personality changes disappear and the person becomes *al-*

most normal again. I say "almost" normal because, although language, memory, and intellectual functioning will improve dramatically, friends and family continue to observe problems in attention and motivation. Unless encouraged, the person with ACC damage tends to remain passive, speaks only when spoken to, and loses the thread of conversations when distracted. In short, the ACC plays a pivotal role in attention, general arousal, and intentional responses to events and people.

Since the ACC communicates with the prefrontal cortex as well as the amygdala and other emotional centers, it serves as the locus where cool rationality and heated emotions converge.

"This overlap provides the ACC with the potential to translate intentions into actions," according to neuroscientist Tom Paus of the department of neurology at McGill University in Montreal. This translation is maximized under conditions of conflict—when your emotions drive you toward one course of action, while your reasoning suggests something else. Neuroscientists know this because of PET and fMRI studies showing the ACC springing into action under situations that require subjects to restrain impulsive actions.

For instance, in what is known as the Stroop Test, different ink colors are used to spell out the names of colors, e.g., the word "green" is printed in red ink. The subject is asked to read aloud the color while ignoring the word (in this example, the correct answer would be red). Since reading ordinarily involves attending to words rather than the color of ink used to print the words, participants experience conflict and a sense of strain. According to fMRI results, the ACC is activated during the Stroop Test, with the greatest activation occurring when the person must respond quickly.

Think of the anterior cingulate cortex as the mediator between the prefrontal cortex, which issues orders ("Concentrate on the color of the word and ignore its meaning"), and the automatic responses that have been built up over the years (reading words for their meaning rather than the color in which they are printed).

The ACC also helps us keep our goals in mind during challenging situations. We know this due to a fascinating study carried out by a team of neuroscientists in Japan. They recorded individual nerve cells in the ACC of monkeys while they completed a series of color discrimination tests. The monkeys were provided with cues that they quickly learned to recognize as harbingers of an imminent food reward. As the clues were given, the monkeys made progressively fewer mistakes in color discrimination. An increase in the firing rate of the cells in the ACC accompanied this enhanced performance. It was as if the monkeys sensed that they were getting closer to receiving their reward and therefore redoubled their efforts. Could this be the correlate of willed control in humans?

Neuroscientists have discovered that ACC neurons fire abnormally in humans over a range of willed motivation and reward disorders, such as Obsessive Compulsive Disorder and drug addiction. Neuroimaging studies of addicts, for instance, show less activity and low cell density in the ACC and associated structures. "It is thus possible that addicted individuals, like others who suffer from hypoactivity in these brain regions, are unable to experience normal affective (emotional) responses to future events or to exert willed control," according to University of Pennsylvania psychologist Laura L. People.

The Amygdala, Anterior Cingulate Cortex, and Free Will

Thanks to the pioneering efforts of researchers like Dean Shibata, William Gehring, Adrian Willoughby, and Laura People, we are now nearing the point where brain imaging can serve as a window into our most private thoughts. In some instances, aspects of our thinking will be revealed that are hidden even from ourselves.

In a famous study conducted by Elizabeth Phelps and her colleagues at New York University, subjects looked at pictures of African Americans and Caucasians. Prior to the testing, the subjects (all of them Caucasian) underwent intensive questioning aimed at eliminating people with an overt anti-Black bias. Despite this, when some of the subjects looked at the pictures of African Americans, their amygdalae became intensely activated— an indication that they were experiencing fear or anxiety.

Those subjects showing amygdala activation then completed standard psychological tests designed to detect racism. The testing revealed that some of the volunteers who denied racist feelings were actively suppressing anti-Black feelings. These were the same people who had previously shown the greatest amygdala activation.

In short, brain activation patterns in one brain area, the amygdala, revealed deeply concealed attitudes about race. While the subjects said one thing, their brains said the opposite. One can easily imagine a not-so-futuristic scenario where those given to embroidering the truth will be unmasked by evidence provided by their own brains.

In the near future neuroscientists will be able to provide "pictures" of brain activity indicating whether someone is telling the truth; whether apparent attitudes and statements are in line with deeply, even unconsciously, held beliefs; and whether emotions are playing a dominant role in the making of a specific decision.

For instance, patients with damage to the anterior cingulate cortex often make impulsive, ill-thought-out life decisions. Absent the input from the ACC, their brain fails to make allowances for the emotions that normally accompany choices about behavior. Thus, the person with ACC damage may respond with impulsive anger and violence to events that a normal person would be able to ignore. That's because the person with ACC damage doesn't experience the normal anxiety and apprehension that accompanies aggressive behavior in most people. Is such a person guilty of a crime if he or she acts impulsively and harms someone due to ACC damage?

With that question we enter some murky waters indeed. Essentially, are our actions freely chosen or are they dependent on forces outside our control? Subjectively, we certainly experience ourselves as free. We can change our mind about something and then change it again. We can even engage in the curious exercise once suggested by Carl Jung as an antidote for indecision: base our decision on the flip of a coin. But even then we remain free to ignore the result of the coin toss and make our decision strictly on impulse.

So if we possess a free will, is it located in the anterior cingulate cortex? Neuroscience is unlikely to ever provide precise answers to questions like that. For one thing, the concept of a free will "center" doesn't mesh with what we already know about how the brain operates. A more useful and answerable question is: Can brain science provide us with standards helpful in establishing whether our decisions are based on reason rather than raw emotion?

We take an important first step toward making freer decisions by learning as much as possible about the automatic and unconscious processing taking place within the emotional circuitry of our brain. And contemporary brain research is providing some powerful insights into the balance between our thoughts and our emotions based on emerging findings on the treatment of emotional disorders like depression.

chapter 7

Cosmetic Psychopharmacology

For centuries the principle treatment for depression was talking to friends. Friends helped by reminding the depressed person that the situation could be worse—he or she wasn't dead, fatally ill, reduced to abject penury, and so on. Later, thanks to Sigmund Freud and others, the treatment of depression came within the province of psychiatrists who formalized "talk" treatment into various kinds of psychotherapy. Finally, starting in the 1940s, biologically oriented psychiatrists discovered that chemicals could relieve depression, thus initiating the era of psychopharmacology—the treatment of mental illnesses with drugs.

Psychopharmacology became prominent in public awareness in the 1990s with the introduction of Prozac and similar drugs called selective serotonin reuptake inhibitors (SSRIs). These drugs relieve depression by raising the level of the neurotransmitter serotonin within the space separating neurons (the synapse) in the brain. (Neurotransmitters function as chemical messengers in the brain. After release from a messenger neuron the neurotransmitter enters the synapse, or gap, that separates it from other neurons. The neurotransmitter then travels across the synapse until reaching a receiver neuron, where it attaches to a specialized receptor and thereby activates that neuron.)

Although psychopharmacology marked a major breakthrough in the approach to depression, a fully satisfying explanation for the various antidepressants' effectiveness remains elusive. For one thing, most of the drugs modify more than a single neuro-transmitter. Some even change the levels of several neurotrans-mitters. What's more, no scientist can explain how or why simply altering the level of one or more neurotransmitters relieves de-pression. Obviously, some basic change in the brain is involved, but what is that change?

A clue emerged from MRI studies of people suffering from depression. The hippocampus—that first important way station in the memory pathway—was found to be smaller than normal in people suffering from depression. Depression, it seems, can lead to atrophy of the hippocampus—one of the reasons that people with depression frequently complain of memory diffi-culties.

Insight into this association came from animal experiments. Depression can be replicated in animals. For instance, if an an-imal is separated from its companions, kept awake, and periodi-cally shocked with an electric grid embedded in the floor of its cage, it will develop a host of endocrine and brain changes associated with depression. This includes atrophy of the hippo-campus similar to the atrophy observed in the MRI of depressed humans.

Microscopic analysis of the "depressed" animal's hippo-campus shows a structural remodeling of brain cells. If a normal neuron looks like a tree in summer with richly thickened foliage, the neuron associated with depression looks like a tree in winter: the dendrites (the receiving "antennae" of the neuron) are short-ened and lack many branches.

Treatment with an antidepressant prevents many of the brain changes associated with depression. In stressed animals that are given an antidepressant, the hippocampal volume increases. Part of this increase comes from increased "branching" of the neurons into a normal "summer" configuration. In addition, new brain cells are produced (the hippocampus is one of the few areas of the brain where neurogenesis—the birth of new cells—occurs). An-tidepressants also prevent several chemical changes brought about by stress-induced depression.

We know that antidepressants prevent stress-induced changes in hippocampal volume, brain metabolism, and cell proliferation. But what is the common mechanism by which antidepressants—many of which affect different neurotransmitters—prevent these changes? One currently popular explanation is that all of the dif-ferent varieties of antidepressants increase the level of neu-rotropins, chemicals within the brain that spur brain growth and development.

In support of this view, one neurotropin, known as brain derived neurotropic factor (BDNF), increases after a depressed person takes an antidepressant. BDNF spurs the growth of new neurons, increases the sprouting of dendritic branches (converting a winter tree into a summer one), and enhances connections between neurons. Current research is aimed at developing additional drugs that function essentially as neurotropins. These new brain drugs hold promise for growing new brain cells or at least stopping or slowing the loss of brain cells and their connections.

Connections and Reactions

Psychopharmacological and genetic research into depression and other neuropsychiatric diseases is likely to advance over the next decade from today's "hit and miss" approach, where a psychiatrist simply tries out a series of drugs until one is finally found to work, to an approach that takes advantage of our increasing knowledge of the biochemical network of the human cell. The first step toward this goal began with the mapping of the human genome, announced on June 26, 2000, at a White House ceremony honoring the two principle genome sequencers, Francis Collins and Craig Ventner. But even the success of the Human Genome Project alone won't provide a fully satisfying explanation of the chemistry of the brain.

To understand why, look at the following typical "line" in the 3 billion letter "text" that constitutes our genome: TAACAT-TACC ATTCAGACCA AATCGAGGCT ACTGGGCATC

TACCATGAAT. The arrangement of these letters—abbreviations of the molecules comprising DNA (adenine, guanine, cytosine, and thymine)—may be the code for a person's hair or eye color, for example. The problem is that in most instances, scientists can't correlate a single line from the genetic "text" with a specific genetic consequence. That's because people simply don't conform to the patterns of genetic inheritance first observed in peas in the 1850s by the monk Gregor Mendel—one gene expressing a single trait. Instead, most of the truly interesting human trait-variations (temperament, intelligence, personality) result from the combinations of many genes rather than single genes. Consequently, the same can be said for most illnesses.

According to physicist Albert-László Barabási, author of *Linked: The New Science of Networks*, "To cure most illnesses, we need to decipher how and when different genes work together, how messages travel within the cell, which reactions are taking place or not in any given moment, and how the effects of a reaction spread along this complex cellular network." What this means is that rather than the gene itself, the ultimate determiner of genetic fate is the network of connections and reactions within the cell. "There are no 'good' genes or 'bad' genes, but only networks that exist at various levels," concluded the historic *Science* article describing the decoding of the human genome.

This important distinction is analogous to my experience with an unassembled outdoor grill manufactured in France that I recently purchased. The grill, consisting of more than a hundred parts, came with a complicated blueprint illustrating the assembly stages. The instructions were in French. Since I'm neither mechanically inclined, skilled at interpreting assembly blueprints, nor proficient in French, I placed an advertisement in the neighborhood paper seeking someone to assemble the grill. But, as I discovered, finding a person knowledgeable enough to assemble the grill wasn't so easy.

A half-dozen experienced grill assemblers looked at everything but declined to undertake the assembly. They said the job was "too complicated." Finally, a bilingual friend who repairs airplane engines for a living offered to take a look. After briefly reading the instructions he sat and quietly studied the blueprint for about half an hour, began to work, and four hours later we were grilling steaks. This was possible only because he could successfully assemble the grill components, understand the blueprint for assembly, and comprehend the French instructions.

Think of the grill components as the genes, the French instructions as the mechanism for conveying genetic information from the genes to the rest of the cell, and the blueprint as the overall diagram for the structure and function of the cell.

A properly functioning cell requires an internal structure along with the energy sources required for the cell's functioning. Barabási compares this cellular energy source to the engine in a car. The engine alone isn't sufficient for your cross-country trip; you also need seats, lights, wheels and suspension, and a network of all the other components that make a car a car. A similar weblike arrangement holds true for a cell. The goal now, rather than focusing on the attributes of specific genes, is for scientists to learn the influences of all the variables in this cellular metabolic web.

"With knowledge of the precise wiring diagram of a cell and the diagnostic tools capable of capturing the strength of the various cellular interactions, doctors in the future could test the response of your cells to a drug before you even take it," writes Barabási in *Linked*.

Your Own Personal Medication

After learning how to understand this "map of life," it will be possible to design specific drugs for specific patients. Your DNA chip—a recently realized concept—will provide the means for developing these truly magical potions.

Basically, a DNA chip is a silicon or glass wafer with an array of tiny holes just large enough to contain short DNA strands, each strand of which contains a single gene. Since the human genome consists of about 30,000 genes, the chip would contain about 30,000 tiny holes—one hole for each gene. Your DNA chip enables scientists to identify the unique messenger RNA (mRNA) created by each of your genes. Here is how this works:

When the DNA in a cell nucleus creates proteins, it does so by copying a gene onto an mRNA molecule, which contains the code or instructions for making a unique protein. The mRNA molecules are, in essence, a precise translation of the gene from which they come. Indeed, the correlation is so exact that each mRNA molecule will stick only to the one hole—one of approximately 30,000—that contains its matching DNA.

Imagine a scientist placing on a DNA chip a cell from a patient suffering from a genetic disease. Only the hole(s) containing the gene or genes responsible for the disease will adhere to the mRNA strands from the patient's cell. A laser reader will then scan each hole in the chip, locating the precise genes containing the DNA that combined itself with the mRNA from the diseased cell.

"With the map of life in hand and with tools such as the recently developed DNA chips that monitor the links between genes, doctors will be able to obtain a detailed list of all molecules and genes affected by a given drug," according to Barabási. How will these new approaches affect brain modification?

Imagine that you suffer from a neuropsychiatric disease, such as bipolar disorder. So far, researchers have identified at least six different chromosomes (4, 6, 11, 13, 15, and 18) that contain genes associated with bipolar disorder. This gene smorgasbord reflects the likelihood that bipolar disorder doesn't result from a single gene but, instead, that several genes may cause the disease by influencing a complex network of biochemical reactions within brain cells. In practical terms, this means that manic depression may result from a dysfunction of different genes in different people. If this is true, then what's needed is a way to identify the gene or genes responsible for *your* bipolar disorder.

Currently your best chance for successful treatment of bipolar disorder involves taking one or more of several available drugs. But the drug that works successfully for another person may have no effect on your bipolar disorder, which is one of the reasons there are so many drugs out there. And even if you're lucky enough to respond to the first drug prescribed by your neuropsychiatrist, neither you nor your doctor will have a clue about the gene or genes responsible for your illness.

Further, since the drug influences many cellular processes in addition to the ones responsible for your illness, you're at an increased risk for side effects and other harmful results. That's because many of the neurotransmitters currently targeted by psychotropic drugs aren't confined to the brain. Serotonin receptors, for instance, line the gastrointestinal track—one of the reasons that patients taking SSRIs like Prozac commonly experience diarrhea and other gastrointestinal complaints.

But even if you're lucky and you don't suffer any serious side effects and your bipolar disorder improves, the drug may nonetheless induce subtle changes in your energy level, appetite, and quality of sleep. And while none of these effects may be serious enough to warrant stopping the drug, neither you nor your doctor would rate the treatment as an unqualified success.

Contrast this with the treatment you could expect within the next decade.

After your initial interview, a sample of your blood (or in some instances spinal fluid) will be withdrawn, and the DNA from your cells will be matched against a DNA chip. This will identify the malfunctioning genes and proteins responsible for your illness.

Next, you will undergo a test known as magnetic resonance spectroscopy (MRS) that provides unique information on the tissue concentration of between 25 and 30 biochemical constituents in distinct areas of your brain. Like MRI, MRS uses a radio frequency (Rf) pulse to disrupt the normal magnetic orientation of protons in the brain areas under investigation. When the pulse is turned off, the protons return to their original magnetic field position. In doing so they emit a radio signal that, when picked up by a receiver coil, forms a snapshot of magnetic resonance in the brain. It is much like what happens when you flick your finger against a compass needle—the needle jiggles and then returns to its original position. The exact timing and extent of the needle's movements, of course, depends on how vigorously you flicked it. Likewise, variations in Rf sequence and magnetic field strengths provide different MRS signals depending on the location and concentration of both neurotransmitters and drugs. A computer then transforms all of this information into a color-coded printout that enables the neuroradiologist to tell at a glance those parts of the brain with the highest concentration of a particular biochemical.

Based on the DNA chip and MRS results, you will then be prescribed a drug specifically designed to correct the unique gene and protein malfunctions responsible for your bipolar disorder. And because of the drug's specificity you won't experience side effects. The drug will have no other actions in your body than correcting the cellular defect or defects underlying your illness.

Just prior to starting the drug you will be injected with a radioisotope that adheres to neurotransmitter receptors in the areas of the brain involved in your bipolar disorder. The color-coded printout that results from a scan of these radioisotopes is, in essence, a brain "fingerprint" that distinguishes the neurotransmitter-receptor pattern of your brain from all others. You'll also undergo an fMRI or magnetic source imaging scan and a magnetic source imaging (MSI) study. The fMRI will show blood flow and metabolic changes related to the activation of various brain structures. The MSI will provide information about electrical currents occurring in the same areas monitored by the fMRI.

After a few weeks on your personal drug, you and your doctor will get together to discuss how you're feeling. Since side effects are unlikely to be an issue, both of you can focus on whether your bipolar symptoms have improved. After the interview you will undergo new scans for comparison with the earlier scans. If the treatment is successful, your improved functioning will be visible on the scans.

To depict visually the signs of your improvement, a radiologist will perform what's called a subtraction study—eliminating from the original scan and the current scan all of the elements that they have in common, revealing in sharp contrast the brain changes that have taken place since you started your treatment.

Imagine looking out a window in your home and taking a picture. Now take the same picture a day or two later. The street, buildings, and other landmarks haven't changed, but the number and arrangement of people walking along the sidewalk will differ from the earlier picture. If you subtract the common elements in the two pictures, you're left only with those features that distinguish the later picture from the earlier one. This is how a subtraction study works with before and after brain images.

With the information provided by subtraction studies and other imaging techniques, your doctor and the radiologist will confer to further refine your diagnosis and treatment. "Most of the people with a scan like that responded to Paxil. I think I'll try that drug for two weeks and then have the patient return to you for a comparison follow-up scan," the neuropsychiatrist may say to his neuroradiologist colleague.

Subsequent MRS and MSI scans will allow a neuropsychiatrist to compare your brain's premedication chemical and electrical patterns with the postmedication patterns. These results can then be used for later comparison after you have stopped taking the medication to inform ongoing treatment.

Lifestyle Drugs

Will the benefits of psychopharmacology and brain imaging technology remain limited to people with diagnosable mental disorders, or will this new and exciting science also have something to offer "normal" people? Indeed, is there a proper use of brain-altering drugs for people who haven't anything wrong with their brains?

At the moment, psychotropic drugs are already being prescribed for people who aren't suffering from emotional disturbances. For instance, modafinil, sold as Provigil (an abbreviation for "promotes vigilance"), was originally developed to treat narcolepsy—a rare disorder marked by the irresistible urge to fall asleep, often under unusual and even dangerous circumstances, such as while driving. Since narcolepsy is rare, one would expect modest sales figures for that drug. Instead, Americans bought $150 million worth of the drug in 2001. Three-quarters of the purchasers didn't have narcolepsy or any diagnosable emotional disturbance at all. In fact, most modafinil users are perfectly normal people taking the drug to counteract drowsiness.

Rather than directing their efforts toward the treatment of sleep disorders like narcolepsy, researchers are studying modafinil's effectiveness in keeping truck drivers awake during long hauls; enhancing the performance of shift workers; maintaining vigilance and performance in soldiers during extended combat missions; and enabling the overworked and the ambitious to work extended hours while going perhaps two or more days without sleep.

Already researchers have learned that healthy people on modafinil can be kept awake for more than two days (54 hours to be exact) without showing any deficit in performance. To put that research into perspective, think back to the last time you missed even one night's sleep and then had to trudge into the office the next day. Your concentration, alertness, efficiency, and emotional control were sorely tested. Wouldn't it have been nice before leaving the house to have taken a drug capable of restoring you to your normal level of functioning? A drug that you could then use to regulate your sleep and wakefulness levels according to *your* wishes. Imagine the decisive edge over your colleagues and competitors such a medicine could provide you.

It's likely that modafinil is only the most recent of a long series of "lifestyle" drugs aimed at people who don't suffer from any diagnosable disorder. In some cases it's becoming increasingly difficult to distinguish between medical problems and lifestyle issues. For instance, if an industrial worker takes a job that requires frequent shift changes and, as a result, experiences difficulty sleeping, along with drowsiness and inefficiency at work, is she suffering from a pathological disorder? Certainly, if we choose to call it a disorder, then it's a strange disorder indeed, since the worker can cure it simply by quitting and taking a job with regular hours.

According to DSM-IV, the official compilation by the American Psychiatric Association of mental and emotional disorders, Circadian Rhythm Sleep Disorder can be diagnosed when a person's body clock (circadian sleep-wake cycle) is normal but there is a "conflict between the pattern of sleep and wakefulness generated by the circadian system and the desired pattern of sleep and wakefulness required by shift work." In other words, lifestyle rather than biology provides the basis for the diagnosis of the disorder.

Shyness or Social Phobia

Social phobia is another lifestyle disorder that people may choose to treat pharmacologically and/or with the latest brain imaging technology. Determining who suffers from social phobia isn't as easy as it might first appear, although specific diagnostic criteria do appear in the DSM-IV. For instance, suppose that you rarely venture out of your house because you fear coming under the scrutiny of others and that you will do almost anything to avoid social contact. If you exhibit such traits, you fulfill the main requirements for the diagnosis of social phobia, and treatment with one of the SSRIs might provide you with relief from your disabling symptoms.

But suppose, instead, that your "social phobia" involves nothing more than a feeling of tension when meeting and talking to strangers. In addition, you feel uptight and nervous when called on to give presentations at work. Yet despite your mild discomfort in these situations and your general preference not to meet too many new people, your professional performance is perfectly acceptable to your supervisors and you maintain regular social contact with a small number of friends. Are you suffering from a disorder, social phobia, or are you, instead, simply a shy person?

Treating shyness as social phobia is only one example of the emerging trend to medicalize and label as disorders mildly impairing but perfectly normal behaviors. This is furthered by direct-to-consumer advertisements of specific drugs used to treat social phobia. These advertisements often imply that only extroverts are really normal. They imply that if you would rather read a book than go to a party, or if you prefer a quiet meal alone or with your spouse over dining at a crowded and noisy restaurant, then you are encouraged to think of yourself as suffering from the "disorder" of social phobia.

A similar medicalization of normal personality traits is underway with depression. The label of depression is bestowed upon anyone who expresses the opinion that, "in general, things aren't going very well," or that "the world is in pretty bad shape." As a result of such medicalization, a person who formerly would have been described as somber, serious, and even philosophical is now redefined as suffering from a mood disorder. And because mood disorders benefit from medication, subtle and sometimes not so subtle pressures may be put on people to take mood-altering drugs.

For instance, people employed in retail sales or other sectors involving a lot of personal contact are encouraged to be perpetually "upbeat." Anything less than a near manic exuberance coupled with a facility for nonstop, often vacuous, social chitchat is taken as an indication of depression. And while it's true that employers have always preferred salespeople with outgoing personalities over their taciturn and introverted counterparts, the preference has become more entrenched now that mood-altering drugs are available. Moreover, even in instances where no disorder is alleged to exist, the normal personality can often be made even "more normal" by the use of these drugs. This, of course, differs from the usage patterns of most other drugs.

An antibiotic, for instance, can bring about a dramatic improvement in someone afflicted with an infectious disease but has little effect in a healthy person. The same selectivity exists for most other drugs—they cure or improve the health of the sick but have little effect on people not afflicted. Not so with psychotropic drugs.

A person with a generally dour and pessimistic personality may "improve" while on one of the newer Prozac-like drugs. He or she may claim to feel better; spouses and friends may even volunteer that they've noticed a lessening of the person's expressions of "gloom and doom." But are such drug takers really better off? Perhaps they are, but can their improvement be attributed to the effect of the drug on a depressive "disorder"? Or has the drug simply brought about a change in what would traditionally be labeled as a pessimistic personality?

Psychiatrists now often diagnose people with pessimistic personalities as suffering from dysthymic disorder. With the exception of feelings of hopelessness (a criteria for dysthymic disorder), many of us sometimes suffer from pessimistic views. This, of course, raises the question of whether dysthymia is actually a disorder or only a normal personality variation. Certainly history is rich with pessimistic personalities who by today's standards would be candidates for mood-altering drugs. What would have been the result of antidepressants on someone like the dour nineteenth-century philosopher Arthur Schopenhauer? Would his improvement have spared him the effort required to write the dark and gloomy *The World as Will and Idea*, which characterizes all human efforts as "doomed to achieve only unhappiness"?

As a result of brain research we have reached the point where powerful chemicals can be used to alter thought and behavior. To that extent, psychopharmacology is a good thing. But those same drugs can be enlisted in the cause of social engineering, and that's not such a good thing. Indeed, the use of medications to improve or alter essentially normal personalities may come at a greater price than many of us realize.

As Simona Folescu wrote in a letter to the editor in the *Washington Post*, "Human development is often stimulated in response to a perceived lack or need. Drugs that mask oppressive social conditions by increasing feelings of self-worth would effectively obscure the need for further social and material progress. A drug-

induced euphoric haze can cloud human awareness to social realities and inure one to the inequality and inhumanity of existing
world conditions."

On a personal level, mood-altering agents can make it too
easy for people to endure situations that they should be changing,
like failed marriages or dead-end jobs. And the drugs can be enlisted to suppress perfectly normal emotions. For instance, patients frequently come to me in my neuropsychiatric practice
asking for tranquilizers to help them get through such experiences as the upcoming funeral of a loved one. They do this because of the belief that any public expressions of grief must be
suppressed in the interest of not making other people uncomfortable.

Are we on the threshold of an age of medication-induced personality transformation? An age where people will take medications to eliminate or reduce life's more discomforting feelings like
sadness, worry, uncertainty, and grief? While it's important not
to allow one's life to be controlled by these feelings, do we want
to eliminate them from our lives entirely? To help answer that
question, let me provide a portrait of what life might be like in
the age of cosmetic psychopharmacology.

Without Any Feelings at All

Our subject is a 54-year-old entrepreneur. Let's call him Ted. As
a result of frequent business travel, Ted's sleep pattern tends to
be irregular and affected by the different time zones he finds

himself in from one day to another. After traveling across several time zones, Ted typically arrives too tired to function efficiently. But Ted has recently solved that problem with modafinil, the sleep drug mentioned earlier, whose exact mode of action remains unknown.

Two weeks prior to the current trip, Ted's younger brother, Jim, died in a car accident. Although Ted was initially upset when hearing the news, he quickly regained his composure by taking a tranquilizer. Three days later he took another tranquilizer so that he could remain calm and composed during the funeral service. Apparently the drug worked, since Ted didn't shed a tear at his brother's funeral. Immediately after the service Ted took a cab to the airport and resumed his business trip.

Now, after two weeks of travel, Ted is starting to find himself thinking of his brother. In the airport in Dallas he suddenly burst into tears "for no reason." But Ted isn't going to give in to any "morbid" responses. Based on his reading he has correctly diagnosed himself as experiencing the onset of depression secondary to an aborted grief reaction. When he reflects briefly on whether he should have acted differently, he ends up regretting nothing. Sure, it might have been nice to cancel his business trip for a few days and make himself available to comfort Jim's widow and children, but what was there to say, really? Jim is gone and nothing is to be gained by morbidly dwelling on that fact.

Ted has in his possession some antidepressants left over from two years ago when he became, in his words, "nonfunctional" and "moped around the house" wondering what "my life was all about." The antidepressant worked and he soon stopped moping and asking himself such questions. Now Ted reasons that the antidepressant should work equally well in regard to his current "fixation" on his brother's death. Ted considers it best to "get over it" and to "move on."

If the antidepressant "works" Ted will soon stop thinking of his brother. He will operate at his maximum efficiency, completely free of grief or any morbid inclinations to dwell on the past. When he gets to his hotel room later he'll accustom himself to the five-hour time change by taking the modafinil. At the end of his business meetings he'll return to his room and take a sleeping pill. And in the morning he'll take his antidepressant and begin his new day *without any feelings at all*.

Although Ted is a composite of several patients I've seen in recent years, he isn't a work of fiction. There are legions of Teds out there fighting internal wars against perfectly normal human emotions and feelings. For patients like him, legitimate uses of the mind- and brain-altering medications are being overtaken by lifestyle applications. But what will become of our capacity for compassion and attachment in a world where such things as grief over the loss of a sibling is considered morbid and best eliminated by taking a drug that will act as quickly as possible? Do we really want brain science to provide us with the chemical means of altering perfectly normal emotional responses?

As you can discern by now, I have reservations about using drugs to manage uncomfortable emotions or low self-esteem. But, to be fair, there is another side to the story. Here is an excellent summary of that position by Tony Boyle in a letter (slightly abbreviated) to the United Kingdom's *Times Literary Supplement*.

"If Prozac—or a superior antidepressant of the future—can promote greater well-being without prohibitive costs, it is difficult to imagine what moral objections could be raised against it. Why would taking Prozac make it pointless to do worthwhile things, like giving to charity or stopping to see an ailing friend in the hospital? No reason to suppose that those on the drug will be less motivated to do such things; nor that Prozac takers fail to get as much satisfaction as the rest of us from such gestures. People often suffer from lack of self-esteem not because of anything they've done, but solely because of neurochemistry. Isn't it generally a boon that antidepressants can favorably alter the chemical balance?"

Whether you consider cosmetic psychopharmacology as a boon or a bust, there is one application that, if you are like most people that I've interviewed, I predict you will be in favor of—drugs to enhance memory.

Memory Enhancement in an Age of "Biobabble"

What is the area of your mental functioning that you would most like to improve? Most people, when asked that question, will pick memory. We all would like to remember more. And memory can be vastly improved by effort and sustained practice, as I've demonstrated in one of my earlier books, *Mozart's Brain and the*

Fighter Pilot. Further, pills already exist that can boost memory in the short term. For instance, amphetamines and other dopamine-boosting drugs have helped students over the years cram huge amounts of information into their brains in short periods of time. Typically, the student takes the drug while studying and again just prior to the test. But while the drugs often work for exams, they are failures as education enhancers. In most cases the students don't retain the learned material after the drugs wear off—a phenomenon called "state dependent learning." In essence, whatever was learned under the influence of a drug becomes lost to subsequent recall unless the person once again comes under the influence of the drug.

But now there is good reason to believe that in the near future neuroscientists will develop a pill capable of substantially enhancing human memory without all of the disadvantages and problems associated with the older drugs. How might such a pill work? To answer that question, consider how normal memories are formed.

Look up from this book and look around you. Select a specific item to be memorized, perhaps the title of another book sitting on a nearby table or shelf. Visual information (the book cover) is combined with language (the title) and that information is funneled into the hippocampus located near the tip of the temporal lobe on each side of your brain.

Within each hippocampus, nerve cells communicate electrochemically. A nerve impulse travels along an outflow track, the axon, until it reaches the end of the line, where it stimulates the release of chemicals, the neurotransmitters, which travel across the synaptical gap separating the transmitting nerve cell from its neighbor, and locks on to special receptors.

Repeated signals from the transmitting neuron increase the concentration of a messenger molecule located within the receiving neuron. Known as cyclic-AMP (c-AMP for short), this chemical relays the nerve signal to the nerve cell nucleus. Finally, via a cascade of chemical reactions, c-AMP activates a memory molecule known as CREB (c-AMP response element binding protein, if you are curious about the acronym). CREB then stimulates the production of new proteins that wend their way back up to the synapse and form your memory of that book title by strengthening the connection between the two neurons. Of course we are talking here not of just one neuron but of thousands, perhaps millions.

Cyclic-AMP and CREB are now the two most popular targets for memory-enhancing drugs. In 1998 one researcher, Eric Kandel of Columbia University in New York City and winner of the 2000 Nobel Prize in Physiology and Medicine, carried out an experiment on aging mice that had lost their ability to navigate their way through a maze. He gave them a drug that blocked the destruction of c-AMP, thus increasing the availability of the messenger chemical. Sure enough, the old mice improved their ability to run the maze.

Another neuroscientist, Tim Tully, is concentrating on CREB. So far Tully and his team at Helicon Therapeutics in Farmingdale, New York, have screened more than 200,000 chemicals in a search for ones that boost CREB and c-AMP. Other researchers are honing in on different brain sites known to be important to conveying information among brain cells. Most are receptor sites—areas where neurotransmitters lock into place and then convey information from the messenger to the receptor cell. Genes are also receiving their fair share of attention. As described earlier, the use of DNA chips makes possible the scanning of thousands of genes in search for hitherto unknown genetic contributors to memory.

In February 2002, Tully predicted to science reporter Robert Langreth that success in the search for a memory enhancing pill is "not an if—it's a when." If this is true, then we should start asking ourselves some questions.

Do we really want to remember just about everything that we've learned and experienced? Most of us would agree that some things are best forgotten. Troubling memories can even lead to emotional disturbances like PTSD. In addition, forgetfulness is an important part of normal mental function. Would you really want to remember every telephone number you have ever dialed? Such an all-encompassing power of recall might even prove painful and frustrating, as it did with many famous memory savants who complained of their "gift" of a perfect memory.

Rather than cluttering our brains with useless memories, most of us would like to improve our ability to remember important things like names, dates, and items related to our work and social life. Moreover, we'd like to recall this information instantaneously, suggesting that we actually favor rapid information retrieval rather than a vast increase in the storage capacities of our brain. And such a preference for rapid processing makes a good deal of sense; since most normal memory failures are time linked, we find ourselves saying, "Wait, wait! Don't tell me."

Think back to your most recent memory "failure." Unless you're suffering from the early stages of Alzheimer's, you didn't forget a block of information (where you were or what you did yesterday) but, instead, experienced a momentary difficulty in retrieving something that was "just on the tip of my tongue." A "memory pill" could be of great help in decreasing the frequency of such momentary failures of retrieval.

A pill to erase memories might also be in our future. The science behind this would involve suppressing c-AMP or CREB or other memory molecules. The pill, taken very soon after a painful event, could block that event from being incorporated into the brain's long-term memory circuits. And since the transfer from short-term to long-term memory can take up to 24 hours, memories could be chemically blocked even if the antimemory pill were taken several hours after a trauma.

But memory-effacing drugs present their own brand of challenges. Traditionally, we've been led to believe that character is formed by facing up to difficulties and overcoming them, rather than avoiding or fleeing from them. What consequences would follow from gulping down an antimemory pill moments after undergoing a distressing experience? Based on my own experience with patients suffering from amnesia, I don't think it would be such a great idea.

Frequently patients who have suffered a concussion in an auto accident complain bitterly to me that they cannot remember the accident. This frustration is particularly marked after accidents in which other people have been killed. "Why do you want to remember something like that?" I used to ask them. "Nature has been kind. Stop trying to remember something that will only give you pain," I would say. But I've discovered that most people want to remember their accident so that they can incorporate it into their life experience. Unrecalled, the accident exists as a kind of foreign body—a psychic splinter—that chafes and irritates.

But I can also think of situations where the distress is so excessive that I wouldn't blame anybody for wanting a memory-erasing pill. I'm thinking now of the rescue workers called to the scene of the Oklahoma City bombing. Most were volunteers with no training that would have prepared them for what they found at the scene. Subsequently, both the rescue workers and the survivors of the bombing experienced the highest recorded prevalence of psychiatric illness among any of the dozen disasters previously researched by the emergency psychiatry unit at Barnes-Jewish Hospital at Washington University in St. Louis. These same researchers predict that as a result of the Twin Towers terrorist attack, psychiatric illness will be "more prevalent, possibly affecting tens of thousands of directly exposed individuals."

If I had been at either of these horrific events, would I have taken a memory-effacing pill if one were available? Frankly, I'm not sure. Would you have chosen to take such a drug? Like many other questions raised by research in the era of the New Brain, neuroscience alone cannot provide wholly satisfying answers.

In summary, our increasing knowledge of the brain is leading to the development of sophisticated experience-altering drugs.

But at the same time, such advances bring about fundamental changes in our self-conception.

Herein lies the conundrum: If we think of ourselves as little more than chemical machines that can be altered by drugs, then what happens to traditional concepts like free will and personal responsibility? While for the most part advances in our understanding of the brain lead to enhancement of, rather than limitations on, our freedom, what will be the overall result if the benefits come at the price of biobabble: people interpreting their experience in chemical terms rather than interpersonal ones?

"I've come to see you in order to get a medication that will increase the amount of serotonin in my synapses," as one patient expressed himself on the occasion of his first office visit for what was clearly a moderately severe clinical depression. In the age of the New Brain it will be necessary to tread carefully, lest we imprison ourselves in concepts that diminish, rather than enhance, our freedom.

chapter 8

Healing the Diseased Brain: New Attempts at Brain Repair

Each year an estimated 730,000 Americans suffer a stroke—an abrupt blockage of the blood supply serving the brain. Strokes can result from the closure of a major blood vessel serving the brain (a thrombosis), the arrival of a blood clot carried through the bloodstream to the brain from someplace else in the body (an embolus), or from a rupture of a blood vessel in the brain (a hemorrhage).

Whatever the cause, a stroke can wreak devastation: paralysis of one side of the body, disturbances in speech, loss of balance and coordination. The exact symptoms depend on the specific parts of the brain deprived of oxygen. Most commonly, a stroke victim loses the ability to use the arm and hand on one side.

To get a feeling for the effects of a stroke, try tying a shoe or buttoning a shirt with only one hand. Not easily done, is it? Yet people who have suffered a stroke are forced to make such one-handed efforts after their brain injury. After a while they may become so accustomed to their paralysis and so dependent on their normally functioning arm that they seem to forget that the paralyzed arm even exists. For instance, the stroke sufferer, when eating, will use the good arm to handle the utensils while keeping the paralyzed arm on his lap. If it becomes necessary to cut meat or butter bread he may ask for help or simply give up in frustration.

Stroke rehabilitation efforts have traditionally focused on enhancing the performance of the nonparalyzed side. But trying to make one limb do the work of two simply doesn't work for many activities. To make matters worse, after a stroke the person often finds it embarrassing and tends to hide or ignore it out of fear that the paralyzed limb can never be the same again.

But this fear reflects an outmoded view of brain development and the possibilities for healing. Neuroscientists once believed that brain development ceased sometime during young adulthood. "It's all down hill after 35" was a typical expression of such pessimism and determinism. But over the past two decades researchers have discovered that brain circuits can be changed for the better or worse, depending on our activities and experiences. We now know that the brain can rewire its circuitry at almost any age.

After a stroke, the brain undergoes a complex pattern of reorganization. Initially, the size of the cortex devoted to the damaged area gets smaller and less excitable. This *injury-related cortical reorganization* is followed by a second stage: *use-dependent cortical reorganization.* In this stage the brain forms new circuits based on increased attempts at use of—that is, mental effort to move—the paralyzed body parts. As a result of these efforts, the area in the cortex serving the injured limb gets bigger. In other words, we encounter here another example of plasticity and the "use it or lose it" formula that applies to the brain on every level and at every age.

Rehabilitation specialists asked themselves, "Could what we know about the brain's ability to change its functional organization be applied to the rehabilitation of stroke patients?" To find out, a group of neuroscientists induced in monkeys a small stroke that impaired the animals' ability to perform precise finger movements of one hand. The experimenters then established a two-part rehabilitation program. The *restraint* portion of the program involved placing the monkey's normal hand in a special jacket. This left the animals with no alternative but to use the impaired hand. During the *training* portion of the rehabilitation the monkeys practiced specific movements with their impaired hand alone. Eventually the monkeys regained nearly normal use of the stroke-damaged hand.

Taking his clue from the monkey study, Edward Taub, a rehabilitation specialist in the department of psychology at the University of Alabama at Birmingham, treats stroke patients with what he calls Constraint Induced (CI) Movement Therapy. Taub constrains the use of the good arm by placing it in a mitt or a sling that the patient wears during 90 percent of his waking hours over a two-week period. During ten of the days, the patient participates in six hours of training of the weak arm per day, practicing functionally important movements such as grasping small objects, using a fork or a spoon, cutting meat with a knife, etc. The goal is to reorganize the brain's circuits and reinstate some of the patient's lost movements by mobilizing the brain's inherent plasticity.

According to Taub, "We believe that damage to the brain from any cause is followed by a period of slow healing. During the early stages of this healing the injured person learns not to try to use the injured part because it doesn't work. We call this learned non-use."

Taub's treatment results in a reversal of learned non-use. Using CI therapy, some individuals with long-standing paralysis of a limb have regained its use in ways that are indistinguishable from prior to the stroke. Measurements actually show growth in the formerly shrunken cortical portion of the brain responsible for the affected limb. This increase in size is thought to result from an increase in the excitability of the neuronal network in the hemisphere damaged by the stroke. In addition, the networks recruit additional neurons to compensate for the damaged area. But why is it necessary to limit the use of the good hand in order for these changes to occur? One explanation comes from a study of kittens.

The Purposive Brain

In the 1960s David Hubel and Torsten Wiesel carried out experiments in which they sewed shut one eye of a newborn kitten. They found that vision in that eye would develop normally if they removed the sutures prior to eight weeks of age. But if the eye remained sutured beyond eight weeks, useful vision in that eye failed to develop. As Hubel explained it, "The closure of one eye produces effects largely as a result of competition. It is as though the neurons from the open eye had somehow taken advantage of their rivals from the closed eye by making more connections. In the competition for space in the visual cortex, the neurons from the open eye had won."

The Hubel-Wiesel kitten experiments provide justification for early operations to remove cataracts from newborns. If too much time elapses, an eye with a cataract will not be capable of providing useful vision because the developing normal eye will have taken over all the cortical circuitry ordinarily shared by both eyes.

A similar situation occurs after a stroke. With the loss of neurons controlling movement and sensation in the paralyzed limb, neurons controlling the other limb establish more connections. As a result, the cortical area serving the normal limb grows at the expense of the cortical area for the paralyzed limb. The resulting imbalance in size and circuitry is even greater if the stroke victim uses the good limb exclusively. But restraining the good limb and using the weak limb corrects the imbalance and restores useful function. The key is to ensure that the undamaged parts of the brain don't overshadow and dominate the impaired parts. "Restrain-Retrain" is the operative phrase.

Despite its recent introduction, Constraint Induced (CI) Movement Therapy is based on one of the oldest and most fundamental principles of brain operation—that the brain is a purposive organ. Its actions are not limited to only a few invariant pathways but can be carried out in a wide variety of ways.

For instance, while I type these words on my word processor, I could just as easily be writing them out by hand (as I did for my first three books); or dictating the words for later transcription; or employing a speech recognition system; or doing no more than simply thinking about what I intend to write at a later time. Each of these activities employs different brain circuits, different muscle groups, and different methods for organization and composition. Moreover, depending on my choice, I unconsciously direct the brain to use some pathways rather than others. Thus, the brain exhibits a marvelous plasticity, with its pathways remaining malleable and subject to change based on life experience and the demands placed on it.

In an experiment carried out by neurophysiologist Mriganka Sur of the Massachusetts Institute of Technology, newborn ferrets underwent an operation where the animals' visual impulses were rerouted to the auditory, rather than the visual, area of their brains. In other words, impulses from the eyes wound up in areas of the brain usually reserved for impulses from the ears.

Within weeks the animals were processing vision in centers of the brain where sound is usually processed. In essence, the animals were seeing the world with brain tissue originally intended to respond to sounds. Sur's experiment indicates that the so-called "centers" for sight, sound, and other sensations are not predetermined but can be altered by experience and necessity.

The Blind and the Musically Impaired

Braille reading provides another example of the brain's plasticity. When reading Braille, blind adults activate not only their sensory and motor cortex in areas serving the "reading" fingers, but also the parietal, temporal, and occipital lobes where vision is processed in sighted persons. Even more intriguing, these same areas are activated in a sighted person during visual imagery—just picturing something in one's mind. With practice in reading Braille, a blind person modifies her brain so that the sensation of touch activates the cortical regions that normally process vision. Her fingers literally become her eyes.

And not every blind person reads Braille the same way. Some readers use one finger, while others use three fingers at a time. These differences in technique not only influence reading efficiency (three-fingered Braille readers are superior readers compared to their one-finger counterparts) but they also bring about distinct changes in brain organization.

"Those who read Braille for several hours a day and use several fingers simultaneously develop a kind of merged, giant, large finger," according to Thomas Elbert, a psychologist and brain researcher at the University of Konstanz in Germany. "They merge information from the different fingertips and send it along a superhighway from the fingertips to the brain."

The merging of the three fingers results in a curious phenomenon. Touching one of the fingers fails to elicit the precise sense of localization that occurs when a sighted person closes his eyes and someone touches the tip of one of his fingers. Elbert believes this results from a fused representation of the blind person's three fingers within the brain instead of a separate sensation for each finger. In essence, since the three fingers work as a unit, the brain fuses and merges their sensations. To get a feeling for how this is experienced try the following experiment.

With your shoes and socks removed and your eyes closed, ask someone to apply a light touch to one of your toes. Then have them touch the tip of one of your fingers. You'll discover that it's easier to identify the finger they touched than the toe. The reason for this discrepancy? There are two reasons, actually. First, the fingers possess a greater number of nerve fibers than the toes. Secondly, since most of the time we're walking with shoes on, all of our toes are stimulated simultaneously as a unit. As a result, the brain represents them as a unit. Each toe can neither be easily distinguished by touch nor in most cases moved separately from the other. While it's easy to move individual fingers without moving others, individual toe movements are almost impossible without practice. It's not impossible, just difficult, and requires special intensive effort. For instance, escape artists engage in hundreds of hours of training to achieve sufficient facility that they can use both hands and toes to untangle knots.

A similar reconfiguration occurs within the brain in response to the training and practice of musicians who play stringed instruments. Magnetoencephalographic recordings reveal among players of stringed instruments substantial cortical reorganization and expansion of the areas responsible for the fingers of the left hand. This occurs because a greater degree of dexterity is required in the left hand (controls changes in pitch) compared to the right hand (controls timing and subtle differences in pressure).

Reconfigurations of the brain's organization in musicians sometimes results in a crippling disability involving loss of control over one or more fingers, usually in one hand. The fingers become cramped and distorted in a posture that neurologists refer to as focal dystonia. For a professional musician the condition can end a career. Musicians may blame themselves and conclude that the condition is based on faulty musical techniques, but recent work using magnetic scanners reveals that the problem resides in the brain, not the hands.

Here's an example. A flutist went for help because she could no longer play her instrument. As she started to play, the fourth and fifth fingers of her left hand assumed a clinched rigid posture. Investigating her problem, the researchers delivered controlled, gentle taps to her fingers and repeated the same procedure with another flutist who had normal use of her fingers. While delivering the taps they measured the resulting electrical and magnetic potentials, the so-called somatosensory-evoked potentials. The principle behind this test is a simple one.

Think back to those occasions when you have accidentally struck your elbow and immediately felt a tingling sensation in the fourth and fifth fingers of your hand. That stimulation of the "funny bone" involved a mild trauma at the point where the ulnar nerve runs through a bony groove at the elbow on its way to the fourth and fifth fingers. At the moment of impact, an impulse from the stimulated ulnar nerve traveled downward to the fourth and fifth fingers and upward from the elbow into the spinal cord.

It continued to wend its way upward until eventually reaching a part of the brain known as the somatosensory cortex.

Each part of the body is represented within the somatosensory cortex, not according to its size but in relation to its importance in everyday life. Thus, in the rat brain a large amount of brain tissue within the somatosensory cortex is dedicated to representing the rat's facial whiskers; in the raccoon brain it's the paws that take up a disproportionate amount of space; in the platypus brain, the bill takes up the most room. In humans, our face and hands occupy much more space than larger body parts like the trunk or legs. This allocation of brain space makes sense because of the importance in our lives of facial expression, speaking, and fine motor control via the hands and fingers. Usually the elbow plays a minor role in our lives and, as a result, we're not conscious of it except when we bump it, setting off the electrical impulse that terminates in the small area of the somatosensory cortex representing the elbow.

Back to our flute players. Based on their understanding of the organization of the somatosensory cortex, researchers electrically stimulated the fingers of a healthy flutist and then recorded the magnetic field changes in the somatosensory area representing the fingers. The recording was normal. But when they stimulated the fingers of the flutist who suffered from dystonia, the magnetic fields were disorganized, enlarged, and formed inconsistent patterns.

Treatment of dystonia uses this knowledge and typically consists of restricting the movement of one or more of the involved fingers. This restriction is combined with a series of exercises using the normal fingers. The goal is to disrupt the brain's fused cortical representation of the fingers that suffer from dystonia. Changing the musician's finger movements converts a dysfunctional pattern into a functional one. Again, this alteration of the brain's circuitry is possible thanks to its plasticity and flexibility.

Other remarkable changes in brain organization occur in the blind. For example, an enhanced "obstacle sense"—the ability to walk without colliding into objects—has long been observed among blind people. Originally attributed to an appreciation of subtle air currents set up by the obstacle and detected by facial skin, a blind person's ability to sense objects several feet away actually results from heightened sensitivity to auditory clues. Better hearing compensates for the loss of vision.

Normally sighted persons can develop a similar sensitivity. With training and effort, blindfolded people can improve their ability to avoid obstacles as the brain reconfigures its organization for sight and sound. This reconfiguration can occur in an extremely short time for the normally sighted person temporarily deprived of vision.

Sensory Substitution

Probably the most amazing example of brain plasticity is *sensory substitution*, where the skin or another body part acts as a receptor of visual information. Sensory substitution is based on a fundamental but counterintuitive rule of brain operation: we see, hear, taste, and smell with the brain rather than with the eyes, ears, taste buds, and nose. Take vision, for instance.

Images aren't conveyed as *images* along a direct path from eye to brain. After passing through the lens of your eye, images progress no further than the retina. Beyond the retina, the brain transforms images into patterned nerve impulses. These nerve impulses are indistinguishable from nerve impulses sent from anywhere else in the body. We know that the brain functions as the final interpreter of whether a nerve impulse is carrying information from the eye or ear. It does this by decoding the nerve impulses coming from the eye and interpreting them in the form of an image.

But decoding and interpretation involve emphasis. Since the brain is a purposive organ, it tries to interpret and makes sense of the world. To do this it emphasizes some aspects of the environment while ignoring others. For instance, you are concentrating just now on the lines on this page and aren't thinking at all of the sounds around you until *now*, as my mention of surrounding sounds brings them to your attention. Your brain ignored those sounds until I turned your attention toward them because your awareness of the sounds detracts from your ability to concentrate on what you're reading. But if a sound is an alarming one, like the siren of a fire truck, your attention will shift instantaneously from the book to the events happening in the street. That happens because, again, the brain is purposive—not simply a recording device—and can easily substitute one sense for another as it seeks to interpret the world.

One of my former editors soundproofed her apartment so

that she wouldn't be disturbed by the noise from the busy New York City streets while reading manuscripts. Now imagine her looking up from her reading one afternoon and seeing a fire truck outside her window. Although she can't hear anything because of the soundproofing, she can infer from the flashing lights on the truck that she should leave the apartment and assess the possible danger of the situation. Deprived of the sound of the sirens, she has to rely on vision, rather than hearing, to provide the first warning of possible danger. And in this instance, vision is less efficient, since the sound of sirens can be heard long before the fire truck comes into view. But the brain will work with whatever information is available to it. Deprived of sound it will rely on sight (and vice versa) or touch, taste, or smell.

"The brain can be taught to extract the same information from trains of nerve impulses from another sensory organ if that organ is carrying information from a substitute 'eye' such as a TV camera," according to Paul Bach-y-Rita of the departments of rehabilitation medicine and biomedical engineering at the University of Wisconsin-Madison.

Think back to the experiment mentioned a little bit ago in which fibers carrying visual information were directed to the auditory area of the brain. If the brain functioned in a machinelike manner, the nerve impulses from the eyes would have been heard as sound—or at least that's how the brain would have tried interpreting them. Instead, the brain reorganized itself so that the hearing cortex came to have the circuits and connections similar to those that mark the visual cortex. Thus, despite the new placement, the brain, thanks to its plasticity, continued to interpret the train of nerve impulses coming to the eye as visual rather than auditory information.

Plasticity can be extended even further in sensory substitution experiments. Bach-y-Rita has developed an apparatus—the tactile vision substitution system (TVSS)—which allows blind people to "see." To use the device, a blind person learns to control a portable TV camera, which functions as an "eye." The image from the camera is transformed by a computer and conveyed to an array of tactile stimulators mounted on the back of the blind subject. As the blind person holds the camera and pans back and forth across an unknown object, small parts of the image stimulate discrete points on the skin surface. With practice, the blind person can recognize the object by "seeing" it via

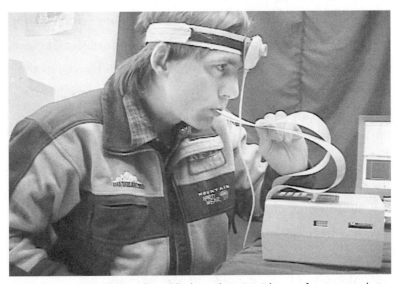

Dr. Paul Bach-y-Rita's device allows blind people to "see" by transforming visual stimulation into tactile stimulation.

the skin stimulation. Here is how Bach-y-Rita explains the process.

"In our sensory substitution studies, the skin serves as the receptor surface to mediate transmission of the information about the artificial receptor (TV camera) to the brain. The skin assumes the role of a relay of the information from the TV camera. The blind person learns with practice to deal with information in context, and thus does not confuse information arriving via the TV camera with information related to normal skin sensations such as tickle or warmth."

At first glance Bach-y-Rita's sensory substitution system seems bizarre. But it's actually only a refinement of what takes place when a blind person walks with the aid of a long cane. While walking with the cane, she perceives a step, a doorway, a curb, a trashcan, a puddle of water; but in the process of perception, the blind person isn't aware of any sensation coming from her hand or her arm that controls the cane. The objects she perceives through the cane (trashcan, curb) are perceived in their correct spatial locations, not to the hand that contains the receptors that initially provide the tactile information.

In essence, the blind person's brain converts into spatial information the tactile information about terrain conveyed from the cane to the touch receptors in the hand. Thanks to this sensory substitution, touch sensations are experienced spatially—the path is either clear or obstructed (keep walking or stop), the pavement remains smooth (keep up the pace), or a curb is imminent (slow down and be prepared to step up or down).

As another way to understand sensory substitution, pick up

a pencil and write a short sentence summarizing today's weather. In order to write the sentence, you'll need to form a mental "snapshot" of how things look outside. And while you're writing the sentence you'll be concentrating on that snapshot, paying no attention to the pressure sensation created by the pencil against your fingers or of the movement of the pencil across the paper. This makes a good deal of sense, since focusing on the weather makes it easier to write well and concentrate on your description of it. Bach-y-Rita suggests this homely analogy as an explanation for how the blind person can use her skin surface as an "eye."

"Just as people can fail to attend to the sensation of pressure of the pencil with which one is writing, while concentrating on the words on the paper resulting from induced pressure on the pencil, so can blind subjects fail to attend to tactile sensations while attending to the information arriving via the camera."

There is one more important point about sensory substitution that applies to both the cane and the pencil examples. The blind person must actively move the cane herself; no one else can do it for her. As tactile information arrives, the brain has to program a motor response to either continue walking while thrusting the hand and cane outward (no obstruction), or stop walking and pull the cane backward (as it makes contact with the trashcan, for instance). Based on additional sensory feedback, a monitoring process repeats itself in hundredth-of-a-second intervals, spatially locating the obstruction via the sensation conveyed by the cane, and then responding by advancing, standing still, or retreating.

Likewise, you will quickly "lose your way" in writing the sentence about the weather if I grasp your hand and put it through the motions of writing the sentence. While I'm controlling your hand you'll have no idea what I'm writing, especially if, as part of our informal experiment, you keep your eyes closed. In fact, you won't be able to confidently identify the sentence by "feel" even if I tell you ahead of time what I'm going to write! Intellectually you'll remember the sentence but will have difficulty correlating each separate word with the separate movements that I'm making with your hand. Test this for yourself by getting someone to move your hand in the writing of a sentence.

Both the movement of the blind person's cane and the pencil in your hand require *active* movement in order for the brain to identify location. The blind person "sees" the curb in front of her because her brain has linked the sensation from the curb (conveyed from cane to arm to brain) and coupled this sensation with the muscle movements and excursions used in tapping the cane against the curb.

Active movement is also requirement for the camera that allows sensory substitution in the blind person. While the skin serves as a "visual" receptor organ, the mobile TV camera provides a means for actively scanning the subject of interest. And for the sensory substitution to work, the camera movement must be under the subject's control at all times. "In the absence of subject-controlled movement (hand and head movements) there is no possibility of identifying the location of the viewed object," according to Bach-y-Rita.

One more way to appreciate the process of sensory substitution is to imagine yourself directing a small camcorder at a child blowing out the candles on his birthday cake. You record the scene from many positions in order to heighten its visual richness. While recording, you "see" the child and the cake in your mind's eye from several different angles, close up (zoom in) and as part of a larger background (zoom out). And while all of your other senses are still operating, you're largely ignoring the touch, taste, smell, and sounds impinging on you in the interest of capturing this very special visual image. In order to facilitate this capture, you employ touch from your hands and muscle movements from your neck, shoulders, and hands in the service of vision, moving the camera into different positions in order to make the best possible recording. A similar situation occurs when the blind person is actively engaged in the movement of the camera: sensations from the skin (the sensory component) and muscle movement (the motor component) come together within the brain as an image. What would ordinarily be *felt* information from the skin is now *seen*.

Refining Sensory Substitution Devices

Results from Bach-y-Rita's system have allowed its users to recognize faces and the experience of "seeing" images sophisticated enough to include perceptions of perspective and depth. But despite this success in transforming tactile information into useful vision, the actual use of a TVSS has turned out to be more diffi-

cult than originally conceived. For one thing, the electrical stimulation via the touch receptors is uncomfortable, and the mechanical gear is bulky, noisy, and requires a lot of electrical power.
Added to that, the overall "weirdness" of the system is generally
off-putting, both to the user and to others. What is needed is a
small, inconspicuous unit that is both effective and cosmetically
acceptable.

Bach-y-Rita came up with just such a system that uses the
tongue as the interface. The tongue is highly mobile, with sensory receptors arranged on its surface, thus allowing for greater
perception of electrical stimulation than is possible even using
the highly sensitive areas at the tips of the fingers.

To understand how a tongue-based system might work, consider a blind person wearing an eyeglass frame containing a
miniature TV camera, a battery power system, and a receiver capable of carrying the televised image. Within the mouth, a dental
retainer holds the receiver and the electrotactile stimulator that
delivers information to the tongue supplied by the camera. In this
substitution system, "vision" provided by the camera image
travels to the tongue instead of to an area on the surface of the
skin.

This tongue-based machine needn't be limited to the blind or
those with physical impairments. Instead, it could potentially
provide access to important sensory information that can't get to
the brain because of a temporary overload in another sensory
organ. A soldier, for instance, could be provided with an infrared
vision system with a tongue interface that would work more ef-

ficiently than simply switching back and forth from normal night vision to infrared night glasses. In addition, visual information reaches the brain faster via the tongue than the eye because of a visual information delay that occurs at the retina.

A pilot outfitted with a tongue-based sensory substitution system could combat visual overload by "reading" his instruments with his tongue while simultaneously keeping his eyes firmly fixed on other features of the flight that require his attention. One study already underway is exploring the feasibility of using a tongue-based system to provide navigational information to Navy SEALs carrying out underwater assignments. And surgeons in the operating room could obtain touch, pressure, and force information by using a tongue-based system coupled with minimally invasive surgical devices. Such a system could even have an impact on recreational games, hyping up a video game to include more than just the visual and auditory sensations. For instance, a participant in a Grand Prix simulation game could experience through the tongue interface the sensation of spinning out of control in a Formula One racing car.

"Given that it is possible to provide information from any device that captures and transforms signals from environmental sensors, it should be possible to develop humanistic intelligence systems to augment and manipulate that sensory information to afford a superior perceptual experience," according to Bach-y-Rita, who suggested the above hypothetical examples. But before sensory substitution devices come into common usage, certain psychological challenges must be overcome.

"The world seems a drab place."

"In our studies with congenitally blind adolescents and adults, the emotional content of the sensory experiences appears to be missing," says Bach-y-Rita. "They are often deeply disappointed when they explore with their camera the face of a wife or girl-friend and discover that, although they can describe details, there is no emotional content to the image."

In one of Bach-y-Rita's early studies with the skin-based tactile sensory substitution system, two blind university students used their cameras to scan *Playboy* centerfolds. "And although they could describe details of the undressed women, it did not have the emotional component that they knew (from conversations with their sighted classmates) that the experience provided for sighted persons," according to Bach-y-Rita. What could explain such a strange finding?

A clue comes from a famous study of a blind man who recovered his sight in 1958 at age 52 after an eye operation. (I wrote in some detail about this man, known as S.B., in 1979 in my first book about the brain, *The Brain: The Last Frontier*.) Briefly, S.B. had lost his sight at ten months old as a result of an infection of his eyes. The operation carried out more than a half-century later removed scar tissue resulting from the infection.

Within months of the operation S.B. was seeing well enough to recognize people and things around him, tell time at a glance from a wall clock, and generally engage in many of the activities carried out by a person with normal vision.

But the most striking change was a dramatic worsening in his personality. In my interview with the psychologist Richard Gregory, who had studied S.B., I learned that before the operation S.B. had been a cheerful, dominant person. The blind S.B. had ridden on motorcycles, confidently stepped from the curb into heavy traffic, even earned a respectable wage as a shoe repairman. But several months after the operation that restored his sight, this self-confident blind man was reduced to a lonely recluse who complained to his wife and Gregory that "the world seems a drab place." On most evenings he never even bothered to turn on the house lights but sat quietly alone in the darkness. At the root of S.B.'s depression was his inability to shift from a lifelong dependency on hearing and touch to his new "gift" of sight.

As with S.B., the subjects in Paul Bach-y-Rita's *Playboy* experiments experienced emotional difficulties. It was as if the areas of their brains responsible for emotions were disconnected from the missing visual sensory channel. Will the emotional experience return after continued use of the sensory substitution device? So far neither Bach-y-Rita nor anyone else has an answer to that question.

S.B.'s experience suggests that although the brain shows great plasticity throughout the lifespan, some limits on its plasticity may exist, especially when the sensory deprivation begins early in life while the brain undergoes its initial structural organization. If so, sensory substitution may be most effective when employed during the early years.

"Sensory substitution begun in infancy will undoubtedly be the most successful," says Bach-y-Rita. "A blind baby should need very little, if any, training, and the emotional considerations will be a part of the development of the child. We are planning to develop a baby pacifier with the entire system, including tiny cameras, battery, electronic circuit, and electrotactile display built into it. The prediction is that babies using such a device will grow as 'sighted' children."

Of course, a blind infant will not develop normal vision as the result of using a sensory substitution device, but the system will aid in the infant's sensory, motor, and cognitive development. "The device would only be used intermittently, since the blind child will also have to live as a successful blind person," according to Bach-y-Rita.

The Sign Translator

Although sensory substitution devices aren't yet commercially available, future applications of the technique promise to take full advantage of the brain's seemingly unlimited plasticity. One example is the Sign Translator, which transforms sign language into text displayed on a small display screen. It is the invention of 17-year-old Ryan Patterson, a senior high school student in Grand Junction, Colorado.

The Sign Translator consists of a golf glove, a small wireless receiver, a display unit, and some basic electronic components. The golf glove is outfitted with sensors that read the positions of the fingers of the deaf person as he or she signs. This information is processed by a software program designed by Patterson that "learns" individual letters by performing calculations based on the distance between the fingers. But since people's hands differ in size, sign language, like any other language, can vary from one person to another. For these reasons, the software program must learn each individual's signing style.

After some practice, finger movements can be translated into text display in about a fifth of a second. Patterson believes that his Sign Translator, with further refinements, should enable a deaf or mute person to communicate directly with normally hearing and speaking people. Thus, for example, when a deaf or mute person encounters a clerk in a store he or she will no longer be required either to write everything out or depend on a speaking person with sign language skills to act as a translator. Further, the Sign Translator can be piggybacked with a digital pager so that the deaf person can communicate almost flawlessly in most situations.

Another sensory substitution device now at the prototype stage aims at helping surgeons recover the sense of touch they often must forego when carrying out certain operations. Historically, surgeons performed their operations via direct "hands on" approaches—opening part of the body, reaching in while observing and feeling their way around, locating the diseased organ, grasping it, palpating it, applying clamps to cut off its blood supply, and then excising it. At each step of the way the surgeon could directly palpate the organs. The coordination of vision and palpation was so highly tuned in some surgeons that they could even predict the pathological diagnosis.

But as a result of changes in surgical technique, this linkage of seeing and feeling isn't always possible. Many operations, such as closed-chest coronary bypass surgery, are carried out with computer and robotic assistance. The surgeon focuses on a monitor linked to a high-resolution camera inserted into the operative field through a jellybean-sized incision.

While staring at the monitor the surgeon manipulates a joystick to guide robotic scalpels and scissors. Since robots don't suffer from tremors or experience fatigue, the operations proceed steadily and take less time. But one important component is missing: the surgeon can no longer directly *feel* the bodily structures. In some cases this loss can have unfavorable consequences. A lung mass detectable on CAT scan may disappear from sight when the chest is opened and the lung collapses. This wasn't a problem when surgeons made larger incisions and directly palpated the area of interest.

What's needed is a bioengineered replacement for the surgeon's palpating fingers. One device, described in the spring of 2002 at a conference on the future of minimal access surgery, is an electronic "fingertip" consisting of a matrix of 64 pressure sensors. After insertion into the operative field, the fingertip glides over the surface of the tissues displayed on the monitor. When the fingertip encounters a mass or lump, greater pressure is exerted on one or more of the sensors. This information is then conveyed to an array of motorized, pinlike sensors applied to the surgeon's finger. A microprocessor "reads" the pressure detected by each sensor at the operative site and conveys this information onto the pins against the surgeon's finger. The brain then makes the ultimate sensory substitution by interpreting the link between sensor and pin the same way it would if the surgeon were actually touching the tissue. When perfected, this device for remote palpation should provide surgeons with the best of two worlds: the speed and precision of robotic surgery and the sensitivity that is only possible when diseased tissues are "felt" as well as seen.

"I see what you're saying."

While such things may initially seem far removed from our understanding of the brain, sensory substitution is a perfectly normal process that takes place all the time. Indeed, our language is filled with examples of the brain's natural predisposition to substitute one sensation for another.

We speak of "loud" neckties that are brighter than we think proper. We "see" what someone is saying. We try to get a "feeling" for what is going on in a social situation or report that something "smells fishy" about another person's behavior and remark that the behavior also leaves us with a "bad taste" in our mouth. This commingling of the senses is most dramatically illustrated in *synesthesia*, a condition in which specific colors are elicited in response to seeing or hearing words, letters, or digits.

Neuroscientists believe synesthesia probably results from the formation of additional synaptic connections between brain regions that are in close proximity. For example, synasthetes often perceive numerals as having a specific color; fives may appear red and twos as green, or vice versa. This color-number synesthesia is less mysterious when you consider that the parts of the brain that process numbers and colors lie adjacent to one another. Similarly, the brain area that processes touch lies adjacent to taste—the probable explanation for *The Man Who Tasted Shapes*, the title of a contemporary book on synesthesia.

Rather than a "strange" and "unnatural" phenomenon, synesthesia may be perfectly normal in the first few months after birth. Neuroscientists speculate that infants experience sound-vision synesthesia in the earliest weeks of life. Later, the synesthetic experience disappears due to pruning (elimination) of the cross-connection between the visual and auditory centers.

"Synesthesia also plays a role in how we use language," according to V. S. Ramachandran, professor and director of the Center for Brain and Cognition at the University of California, San Diego. Ramachandran, who is an international expert on synesthesia, suggested to me that language and even alphabets employ universal synesthetic rules. For instance, imagine a two-letter alphabet with one round-shaped letter and the other letter resembling a pointed star. Which letter would you guess might be called "booba" and which letter called "kiki"?

"Across native languages, 99 percent of people will select 'booba' for the round letter and 'kiki' for the star-shaped letter. That's because the mouth assumes a rounded contour when saying 'booba' and thus mimics the round contour of the letter; 'kiki,' in contrast, has a sharp verbal edge to it that mimics the sharpness of the pointed star," explains Ramachandran.

Another intriguing variant of synesthesia is called *synkinesia*, in which one part of the body mimics the actions of another. For example, tongue and lip movements may mimic hand movements, as when a person clenches and unclenches his teeth while using a pair of scissors or makes clicking noises while snapping the fingers in rhythmic accompaniment to a song.

Two principles seem to be at work in both synesthesia and sensory substitution. First, since the brain is a dynamic organ that responds in milliseconds, a sensitivity shift from seeing to hearing or seeing to feeling can rapidly occur when vision is temporarily obscured. In addition, the brain's response to perceptions differs according to the circumstances. Responses to dark and dim objects are faster and more accurate when they are accompanied by a low, rather than a high-pitched, sound. And responses to light and bright objects occur more quickly when accompanied by a high, rather than a low-pitched, sound. In psychological tests where participants are asked to associate high and low frequency tones with white or black colors, most people associate the high-frequency tones with white and the low frequency tones with black. Other associations exist between pitch and lightness, pitch and brightness, loudness and brightness, and pitch and shape.

Think of the brain as synthesizing various sensations into a totality. The brain uses one sensory channel to extract information relevant to another channel and then integrates it. Scientists describe this process as cross-modal matching. For example, a man who is looking at an object on a screen can use his hands to select the same object from an array of objects hidden from his sight by a curtain. Infants can successfully make such distinctions, and adults do it daily when reaching into a pocket or purse and extracting a quarter instead of a dime by touch alone.

But there are some interesting limitations on cross-sensory

transfer that tell us a lot about how the brain is organized. In brain damage to the parietal lobes, for instance, the brain loses the ability to recognize and transform something felt into something seen. In one of the tests I carry out to assess parietal lobe function, I ask the patient to close his eyes while I use a caliper to form a letter on his palm. In the absence of parietal damage a person has no difficulty mentally transforming a felt letter into a visually observed one. Try it yourself. Have someone gently form a letter on your palm with the blunt end of a pencil while you close your eyes or look away. You should experience no difficulty rapidly identifying the letter. But then try something trickier.

Trace the form of the letter *b* on a friend's forehead. There's a good chance he'll report it as *d*. If you trace *b* on the back of his head he'll more than likely get it right. Why this difference? In most cases a person evaluates letters on the basis of what the letter would look like if seen. Looking straight ahead from the perspective of one's forehead, an inscribed *b* looks like *d*. But when formed from the rear—a direction from which the subject cannot see directly—the tendency is to mentally transpose oneself to the position of the person forming the letter and thus *b* is correctly identified.

To appreciate the difference, use your forefinger to draw the letter *b* on the back of your head. No problem, right? It feels like a *b* and mentally you can "see" it as a *b*. Someone standing beside you would see it the same way. Now draw *b* on your forehead. Most people draw the letter as they see it from their perspective of looking face forward. But when seen from the point of view of someone standing in front and looking at the forehead, the letter is *d* not *b*. If you still have difficulty on this point, place a piece of paper on your forehead and with a pen or pencil actually write the letter *b*. Then look at the letter you've written on the paper. What you wrote as a *b* turns out actually to be a *d*. And if you attempt to circumvent this by drawing the downward vertical line followed by a loop to the left you have drawn a *d* rather than the *b*.

My point with these exercises is to show that there are some minor but fascinating limitations on plasticity imposed by the brain's organization. In most instances these limitations aren't functionally impairing in any way.

Imaging Pure Thought

With these findings on brain plasticity, neuroscientists are coming to a new understanding of the brain's capacity for change. Rather than requiring years, months, or weeks, some of the brain's functional organization can change in minutes, perhaps even seconds. To personally experience your brain's ability to fluctuate between different neural states, stare at the drawing on page 178. While the optical information entering your brain remains the same, your perceptual interpretation will shift back and forth. One moment you see the star located at the back of the box—the next moment your brain undergoes a shift in perceptual interpretation and sees the star at the front of the box. And however hard you try, you cannot see the star in both positions at the same time—a phenomenon referred to as bistability.

A similar process occurs in so-called epiphany experiences, marked by sudden insight about a person or situation. For instance, a sudden insight may reveal a friend to truly be an antagonist, or a seemingly promising business opportunity may be revealed to be a scam. Of course, such insights might turn out to be wrong, having to do with a separate issue about the brain's capacity for self-deception. The point here is that these judgments can occur in seconds. That's because the brain, when faced with ambiguous situations, may fluctuate rapidly between different, sometimes contradictory or incompatible, neural states.

We often cannot exert conscious control over our brain responses, as shown by experimental research using PET scans. For example, imagine a situation where I flash a series of words on a screen and ask you to pay attention only to certain features of the words rather than to read them—whether certain letters are reversed or upside down, for example. Despite your best efforts to

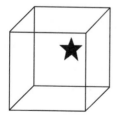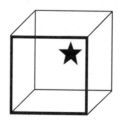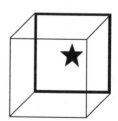

Stare at this figure and notice how the star moves from the top of the box to the back of the box to understand how the brain can rapidly change its functional organization.

concentrate on the letters without reading the words, the process of looking at the words will activate the language areas of your brain. Neuroscientists refer to such a response as hardwired. Presumably an association exists in the brain between words and the activation of certain areas mediating language-related functions, including the appreciation of meaning. This *obligatory activation* occurs even when determined efforts are made to ignore word meaning.

Such a linkage between words and the activation of language areas makes a good deal of sense outside of the laboratory. It would be difficult to imagine where it would be advantageous to look at a word and see it only as a string of letters. Indeed, among individuals who have learned to read, the ability to recognize letters but not words occurs only as the result of certain types of brain damage.

Exercising one's imagination provides another example of obligatory activation. For example, imagining or visualizing a complex or skilled movement can help improve performance, as musicians and athletes can testify. PET scans show that the brain areas involved in such motor imagery surround the areas that are activated when the movement is actually made. The sequence goes like this.

Initially those areas involved in planning the movement are activated, followed by activation of the appropriate motor areas that carry out the action. PET scans make it possible to view brain activity associated with imagining a movement separately from brain activity associated with performing it, by eliminating from the picture those areas activated during the actual movement. "Such simple experiments illustrate dramatically that it is possible to image brain activity associated with 'pure thought,'" according to Richard Frackowiak, professor of neurology at the Institute of Neurology in London.

In another intriguing experiment, researchers used a magnetoencephalogram (MEG) to compare the activation in the motor cortex of pianists compared to non-pianists. Pianists frequently assert that their fingers involuntarily move while listening to a familiar musical composition. The MEG findings not only confirmed this but specified the exact brain areas involved: the cortex on the side opposite to the hand playing the specific notes. The MEG could even distinguish brain activity associated with notes preferably played by the thumb and the little finger. Such an experiment provides an outward, objective demonstration of an internal mental process. It dissolves the previously firm boundary between inside and outside, mental and physical, voluntary and involuntary. The pianists in the experiment didn't consciously will the movements of their fingers; the movements occurred irrespective of their wishes!

In an even more intriguing experiment aimed at reading the thoughts of a person by using observation of brain activation patterns, subjects are provided with pictures of either faces or landscapes. Later they are asked to mentally imagine the faces or the places they had previously seen. Data coders (researchers who interpret data) could accurately assign on 85 percent of the trials whether a subject was in the "faces" or "places" group, simply by looking at his or her fMRI scan. They could do this because different parts of the brain are activated when looking at faces or places (the fusiform area for faces; the parahippocampal area for places).

"We can actually tell, on a moment-by-moment basis, what an individual is thinking about, by measuring brain activity, rather than having the person tell us," according to cognitive neuroscientist Kathleen O'Craven, who led the study at Massachusetts General Hospital in Boston. "We may someday be able to get a glimpse of what a person is able to comprehend."

The Triumph of Phonics

One promising avenue toward achieving O'Craven's goal involves recent brain findings about reading and writing. Imagine yourself riding along in your car listening to an audio recording of the latest Stephen King novel. When you get to your office you're so intrigued with what's happening in the story that you sneak a few minutes to take out a written copy of the novel and, picking up at the point where you stopped listening in the car, you read on to the end of the chapter. While this shift from listening to reading proceeds seamlessly, your brain processes these two different activities quite differently.

"The brain constructs a different message for reading and listening," according to Carnegie Mellon University psychology professor Marcel Just. "Listening to an audiobook leaves a different set of memories than reading does. A newscast heard on the radio is processed differently from the same words read in a newspaper."

Just arrived at this conclusion by using fMRI to measure the activity in 20,000 tiny brain areas, recorded at three-second intervals, while a person either read or listened to the same material. From these measurements, Just and his colleagues created colored visual maps illustrating the moment-to-moment changes. Two surprises emerged.

First, the right hemisphere was not as active as expected while reading—the first suggestion of qualitative differences in comprehension between reading versus listening. Second, listening resulted in an increased activity in a left hemisphere region called the pars triangularis (the triangular section). This is a component of one of the speech and language processing areas, named after the nineteenth-century neurologist Paul Broca, called Broca's area. It also springs into action whenever it is necessary to keep verbal information in mind, such as when a person listens to a list of requested items prior to setting off for the supermarket.

Just's findings make sense when you consider the differences between reading and listening to the same material. Written language is easier to process and remember because the characters making up the sentences can be reread if necessary. Spoken language, in contrast, is evanescent, with each sound suspended in the air for only a fraction of a second. As a result, the brain must store the initial words of the spoken sentence and then mentally link them with subsequent words in order to come up with the sense of the sentence. This requires the smooth operation of a mental playback loop (called the articulatory-phonological loop) that allows someone to continue to "hear" in their head the first part of a sentence while simultaneously listening to the later parts. This is carried out by Broca's area, which explains the enhanced activity observed in this region when a person listens instead of reads.

The fMRI findings don't imply that reading information is better than listening to it or vice versa. "It depends on the person, the content of what's read, and purpose of the reading," according to Just. "Some people are more skilled at reading or listening and typically prefer their more skilled activity whenever possible. It may be that because of their experience and biology they are more comfortable in listening or reading."

It may soon be possible to understand reading/listening preferences from a neurological basis. Recordings using fMRI or other technologies will enable neuroscientists to map the individual differences that distinguish one person's reading and listening skills compared to the average. Already the fMRI studies have shed considerable light on a longstanding and often contentious debate about how children should be taught to read.

On one side of the debate are those favoring whole-word instruction (also referred to as the "look-say" method). A beginning reader taught by this method looks at books and tries to guess the meaning of unfamiliar words by relying on such things as the context of the sentences or content of the illustrations. Little attempt is made to teach the child the relationship between letters and the sounds associated with them (phonemes). Instead, whole-word instructors believe that the phonemes will be learned incidentally through repeated exposure to interesting and captivating stories.

Phoneme-based instruction (phonics), in contrast, teaches the child from the beginning how to use his or her knowledge of the alphabet to sound out words. This reading instruction technique emphasizes the correspondence between spoken sounds and the letters that represent them.

Which of the two methods corresponds most closely to what happens in the brain during reading? Until recently there was no way of conclusively proving one method superior to the other. But recent fMRI studies have largely come down in favor of phonics. It turns out that even highly skilled readers silently sound out words, especially unfamiliar words, while reading. Functional MRI activation of the lip and mouth areas in the primary motor cortex accompanies this silent reading.

Along with insights into the process of normal reading, neuroscience is also providing early detection of dyslexia—a persistent difficulty in acquiring word-reading skill. Dyslexia affects from 5 to 10 percent of school-aged children and frequently persists into adulthood. Dyslexia is more than simple difficulty in reading. Dyslexics suffer from defects in what is known as phonological competence, which means they have problems in rapidly recognizing and naming objects or numbers. Dyslexics also may not be able to hold a word in memory well enough to link the beginning and ending of the sentence.

Dyslexics learn new words more slowly, and learning a foreign language is a particularly agonizing and painfully slow process. Even when reading simple words, the dyslexic's impaired phonological skills coexist with abnormal brain activation patterns. According to one expert on dyslexia, "The dyslexic brain is not able to mould connections between the sight, sound, and meaning of a word as efficiently as other brains."

Although dyslexics, by definition, suffer from impaired reading ability, it's sometimes difficult distinguishing a dyslexic from someone who is merely a slow reader. For this reason, reading difficulties alone aren't always reliable indicators of dyslexia. Neuroscientists have therefore concentrated on imaging the brain during reading. Using fMRI scans, they have found in dyslexics abnormal patterns of brain activity, predominately in the left hemisphere and especially the left temporal, parietal, and inferior frontal lobes.

As I write this I have before me the brain activation profiles of eight severely dyslexic children as they attempted to decide whether two pseudowords rhymed—for example, "yoat" and "wote." If the words rhymed the children raised their index finger; if the words didn't rhyme, they didn't take any action. The profiles consist of colored pictures based on magnetic source imaging (MSI), which directly measures electrical currents in clusters of neurons during reading and provides a real-time, spatiotemporal map of brain activity.

The profiles show brain activity before and after 80 hours of intensive remedial reading instruction, with activation patterns of both the right and left sides of the dyslexic children's brains. In all eight profiles, the left temporoparietal areas show decreased activity compared to profiles of children who read normally. But in profiles taken after the 80 hours of reading instruction, the left temporoparietal activation is normal.

"These findings suggest that the deficit in functional brain organization underlying dyslexia can be reversed after intense intervention lasting as little as two months," according to the doctors who carried out the studies at the University of Texas Health Science Center at Houston. But the studies also point to something even more exciting.

"It appears that although dyslexia has a demonstrable neurologic basis, it is not a neurologic disease. Rather, word-reading difficulties most likely represent variations in normal development that can be reversed by means of reading intervention targeting phonologic processing and decoding skills," according to the Houston researchers.

In short, remedial programs can successfully reverse dyslexia as long as the programs target the underlying problem—the dyslexic's difficulty in grasping the correspondence between letters and phonemes. This approach, known as the alphabetic principle, forms the structural underpinnings for teaching programs aimed at developing phonemic decoding skills.

"When successful intervention occurs, our study suggests that neural systems are altered and that these neural systems are much more plastic than previously believed," the Houston researchers concluded.

While teachers have long championed remedial reading programs as correctives for dyslexia, they had no scientific proof that the programs changed the brain. Now, thanks to the MSI study, they have that proof. Even more intriguing, the study shows that a brain "abnormality" may actually be a variation of normal, rather than something permanent and unalterable.

Basically, the goal now is to employ neuroimaging techniques to illustrate in vivid pictures the heretofore-invisible operations of the human brain. In the process of developing such techniques, neuroscientists are making some fundamental discoveries about who we are and what we do.

chapter 9

The New Brain

In this last chapter of the book I want to look forward and mention some of the applications of neuroscience that are likely to occur during the next decade. Many of these advances will be based on new developments in imaging.

In the coming years we will be able to study mental operations in "real time," as imaging technologies provide moment-by-moment measurements of what's happening in the brain. These technologies promise to reveal the brain processes underlying learning, language, emotional experience, and thinking. Here are some of the imaging technologies recently developed that can already do that.

Magnetoencephalography records the magnetic fields produced by the human brain. These fields are extremely small, nearly a billion-fold smaller than the magnetic fields produced by the earth and other environmental magnetic sources.

In addition to recording the brain's inherent magnetic activity, neuroscientists can also use magnets to alter brain function, through something known as transcranial magnetic stimulation (TMS). This method influences the magnetic fields of neurons both at the point of stimulation and at some distance away. For instance, repetitive high frequency stimulation (rTMS)—defined as a stimulation rate of more than five magnetic pulses per second—increases excitability of the cerebral cortex and is a promising treatment for depression.

Low frequency rTMS at a stimulation rate of approximately one per second decreases cortical excitability and therefore may prove helpful in illnesses marked by cortical hyperexcitability, such as epilepsy and the auditory hallucinations associated with schizophrenia.

TMS has already been found to be effective in speeding up a normal person's mental agility and responsiveness. In one experiment, cognitive neuroscientist Jordan Grafman tested volunteers for their ability to solve geometric analogy puzzles. When shown a group of colored geometric shapes, the volunteers were challenged to pick out an analogous pattern using other sets of shapes. While the subjects performed these tests, parts of their prefrontal lobes (along with some other areas) increased in activity. Grafman found that after administering rTMS to the prefrontal areas the subjects solved the analogy puzzles quicker, but no enhancement occurred if he applied the rTMS to other brain areas or if he only pretended to stimulate the prefrontal lobes.

While Grafman isn't sure why rTMS to the prefrontal lobes works, he speculates that it raises the baseline level of activity to the point where prefrontal neurons have an easier time coming up with a problem solving strategy. Or the prefrontal stimulation might simply enhance concentration and focus. In either case, rTMS might prove helpful to individuals in conditions of severe time limitation, such as an airplane pilot faced with an emergency or a neurosurgeon faced with a particularly challenging surgical dilemma. Self-activation of a brief "burst" of rTMS might prove lifesaving in both instances. High frequency stimulation might also hasten rehabilitation after brain injuries.

The Dynamic Brain Atlas is another exciting future technology. Created by scientists at King's College London; Imperial College of Science, Technology, and Medicine in London; and Oxford University, the Atlas provides an individualized snapshot of a person's brain that is far more accurate and sophisticated than anything else available. After entering data from an individual's brain scan into a networked personal computer, a doctor can compare that scan with a dynamically generated brain atlas customized to her patient.

Much like an Internet search, the doctor enters specific information about the patient, such as age, sex, and general health. Computers from around the world then retrieve images of people's brains that have similar characteristics. The image of the patient's brain is then matched to a customized brain atlas containing the range of normal sizes and shapes based on images of the brains of thousands of similar people.

"Computing power is used to match each reference image in order to make this customized brain atlas," according to Derek Hill, a radiologist at King's College. "After a few seconds the doctor can see the patient images alongside the brain atlas, or can overlay pictures from the atlas onto images from the patient, thus pinpointing those regions of the brain that are abnormal."

So far the Dynamic Brain Atlas is being used to identify subtle brain abnormalities that are beyond the resolving powers of more conventional technologies like MRI and fMRI. But the technique, or something like it, could conceivably be used to compare a specific brain with Atlas-derived images of other normal brains drawn from people of the same age, sex, and general background.

Prior to the development of tools like the Dynamic Brain Atlas, neuroscientists had no way of comparing images of an individual brain with images of the brains of huge numbers of other people of similar background. But in the near future neuroscientists should be able to draw upon images of millions of brains and use that information to prepare a customized brain "map" of any given individual.

We know that no two brains—not even the brains of identical twins who share the same genes—are exactly the same. But so far there hasn't been any way of demonstrating those differences. An Atlas will display identifying characteristics of your brain that are more specific than your fingerprints.

Brain Changes over a Lifetime

But the brain not only differs from person to person; it also differs in the same person during life's developmental stages: infancy, childhood, adolescence, adulthood, and old age. For instance, your infant brain was quite different from your adult brain as you read this book. (I describe the changes undergone by the brain during life's developmental stages in an earlier book, *The Secret Life of the Brain*.) Thanks to fMRIs it's now possible to observe the differences in brain function when individuals of varying ages perform the same actions.

For instance, recent research reveals that the brains of adults and school-age children use different circuits when processing single words. In one study, researchers from Washington University in St. Louis compared the brain changes in a group of 7- to 10-year-olds with a group of adults who were asked to come up with a word that either rhymed with a given word or was the opposite of the word—for instance, the participants were to answer "short" when given the word "tall." In the interests of fairness to both the children and adults, the researchers selected words from a first-grade vocabulary list.

While the children and the adults performed about equally well in generating words, their fMRI patterns differed in two areas: the left frontal region and an area just outside the primary visual area in the left cortex. The extravisual area showed more robust activity in children; the left frontal area lit up more in adults. In short, the brains of children and adults process language by employing different circuits. As a child matures, he or she shifts toward the adult pattern. At the moment, no one knows when this takes place. But the answer should soon be forthcoming since fMRIs, as opposed to PET scans or other technologies requiring isotopes or X rays, can safely be repeated multiple times.

This study is typical of the kind of exploration we can expect to see over the next decade as we seek to understand how the brain changes over a lifespan.

Turning Thinking into Movement— Without the Body

Ethical considerations preclude studying the brain directly by exposing human volunteers to the risks inherent in opening the skull. But, right or wrong, less stringent limitations presently apply to neuroscientists working with animals.

In a typical experiment, a recording electrode is guided into the brain of a monkey, resting in proximity to a small group of neurons. Next, a sound or visual image is shown to the monkey, and a recording is made from those neurons that fire in rapid response to the auditory or visual stimulus. One neuron is then selected and recorded as it fires in response to repeated stimulation by the sound or image. The recordings provide researchers with a millisecond-by-millisecond snapshot of the firing rate of the neuron. Most important, researchers have noticed that the recordings change according to the psychological state of the monkey.

For example, if the animal is trained to direct its attention to sound stimuli, the firing of the cell in the auditory area is enhanced. If attention is directed to another stimulus (light instead of sound) the firing rate in the auditory cell declines. In other words, paying attention amplifies activity in those brain cells devoted to analyzing a particular stimulus. A similar process occurs in our own brains. For instance, looking at a word on a computer screen activates our visual areas. But if, instead of merely viewing the word we try to think of uses for the word or come up with another word in response, other areas become active, especially the anterior cingulate and parietal cortices. Both of these areas, it's turning out, are important in tasks requiring focus and attention.

For instance, as you read this book your attention will vary from moment to moment, depending on your interest and familiarity with the material. The more interested you are, the more your attention is "captured" and the more activated your parietal cortex and anterior cingulate cortex. And when you encounter something that especially intrigues you, these structures are at their most active. Indeed, a PET scan or fMRI scan of your brain while reading will indicate your motivation and interest to keep reading. Such linkages between psychological processes and brain activation are making possible some highly intriguing correlations.

Consider what happens when you reach the end of a page and turn to the next page. Your prefrontal cortex, along with a highly distributed network of neurons throughout your brain, instantaneously draw up a "model" of the situation that includes such factors as the distance from your hand to the page, the movement required of your fingers, the exact amount of pressure and force needed to turn the page, etc. Recordings from your brain would show increased electrical and metabolic activity (use of glucose, sugar, and oxygen) in the cells responsible for the movements.

Now, imagine simply staring at the page while *thinking about* turning it. Many of the same brain areas would be active, except for those areas concerned with actual muscle movement. In other words, the brain prepares for muscle movement milliseconds before the movement actually takes place. It formulates a "motor program" of movement based on a representation of the movement formulated principally in the frontal and prefrontal cortices. From here the program is transferred to the motor cortex, where the movements are actually carried out. Recently, researchers have tapped into areas of the motor cortex responsible for planning the directed muscle movements involved in actions.

In one experiment, a tiny array of 100 miniature electrodes, together about the size of a fingernail, is implanted in a monkey's motor cortex. Recordings from the motor cortex are made as the monkey plays an electronic pinball game by manipulating a joystick. The recorded signals are then used to develop a program that pairs the electrical firing rates of the motor neurons with the resulting trajectories of the monkey's arm as it moves the joystick. After dozens of trials carried out over several months, with orange juice offered as a reward for successful performance, the monkey learns to move the cursor just by thinking.

"We have tapped into an area of the brain that is conceiving of where you want to put your hand," according to John P. Donoghue, a professor of neuroscience who directs the research at Brown University in Providence, Rhode Island. According to Donoghue, only a few minutes of recording are required to create a model that can interpret the brain signals that instruct specific movements.

But the motor cortex is not the only area serving as the source of instruction for controlled movement. Daniella Meeker of the

California Institute of Technology in Pasadena has taught monkeys to move a cursor on a computer screen by activating the parietal cortex. Meeker starts by finding a small cluster of cells— sometimes as few as 16—in the parietal cortex that become active just prior to the monkey reaching out to move the cursor. She then implants sensitive electrodes in their parietal cortex after they have been trained to play a simple video game that involves using their hands to touch dots on a touch-sensitive computer screen.

"After the monkey has done this several times, we are able to determine for a particular neuron the different patterns of electrical activity when the animal is planning to reach in different directions," says Meeker.

Next, Meeker trains the monkey not to move its hand but simply "think about" doing so. A computer program linked to the implanted electrodes provides the basis for monitoring the monkey's thinking by tracking increases in activity in the parietal neurons. The computer then moves the cursor on the screen in the direction the monkey would have moved it if using its hand. "The change in electrical brain activity indicates the monkey is planning to reach toward the screen," explains to Meeker.

After some practice, the monkey becomes reluctant to move his arm or hand once the cursor is introduced into the game, according to Meeker. Instead, the monkey prefers moving the cursor merely by thinking about doing it.

Could such programs be made to work in humans? The answer is a qualified yes. In one current application, Philip

Kennedy, director of Neural Signals, a research and development company in Atlanta, is training a paralyzed man to move the cursor on a computer screen merely by thinking about it. Future applications may involve replacing the cursor with a robotic arm that carries out specific instructions to perform the work of a real arm. Such developments may be furthered by implanting chips directly into the brain, thus freeing a person from the need to maintain a constant link with a machine.

Roborat

At the moment, neuroscientists are exploring in animals the outer limits of brain-machine interactions. One particularly intriguing example involves the rat—a creature that just about everybody since the dawn of civilization has learned to fear and hate.

One person who neither hates nor fears them is John K. Chapin, a physiologist at the State University of New York Downstate Medical Center in Brooklyn. In May 2002 Chapin announced in the journal *Nature* the arrival of the first remote-controlled living robot—a rat outfitted with a tiny backpack containing an antenna for receiving radio signals, a small microprocessor for delivering electrical pulses to the rat's brain, and a miniaturized video camera for delivering the equivalent of a rat's eye view of the world.

Whether you believe the world is improved with the advent of the rat robot, a.k.a. "roborat," you can't help but admire Chapin's ingenuity and originality. He implanted three wires into the rat's brain. Two of the wires stimulate brain areas that detect touch to either the rat's right or left whiskers. The third wire enters the pleasure center in the rat's brain known as the medial forebrain bundle (MFB).

Chapin and his coworkers trained the rats to turn toward the side of the electrical stimulation by rewarding them with a pulse in the pleasure center when they made the correct choice. After only 10 days of this conditioning, the rats could wend their way through almost any landscape within the range of the radio transmitter (550 yards) in response to the directions of a technician issuing commands from a laptop computer. The rats could even maneuver their way through an obstacle course requiring them to climb a ladder, cross over a narrow ledge, descend a flight of steps, pass under an arch, and descend a steep ramp.

Despite claims that the hybrid creatures could be used for rescue missions, I seriously doubt that this "Dr. Frankenstein meets Dr. Strangelove" technology will be ready for primetime anytime soon. Imagine yourself trapped beneath some debris awaiting the arrival of your rescuer. Would you prefer to encounter a St. Bernard with the traditional tiny keg of brandy around its neck or a rat with a weird-looking camera apparatus clawing its way toward you in the darkness?

Predicting Human Behavior

Within the next few years, look for increasing discoveries about genetic contributions to specific cognitive abilities, such as intelligence, attention, memory, language, and physical performance. This new frontier of brain research will be based on what geneticists refer to as polymorphisms—one or more variations of a gene caused by a difference in a single amino acid at one site on the gene. And sometimes the variation in a single amino acid can make a practical difference. For instance, if you like Indian food but your spouse doesn't, that distinction may result from the difference of a single amino acid. Because you possess one allele (the technical term for such variations), you find spicy food enjoyable; if you possessed another allele, you may find that spices "burn" your tongue. Differences in alleles sometimes lead to less trivial consequences.

The enzyme catechol O-methyltransferase (COMT), which breaks down the neurotransmitter dopamine into its component chemicals, exists in two forms. One form contains the amino acid methionine (meth), while the other contains the amino acid valine (val). And practical consequences follow from whether a person's COMT contains the meth or val allele. People with the meth allele perform significantly better on a test of frontal lobe functioning than people with the val allele. Thus, knowledge of a person's COMT meth/val status provides a basis for predicting how well that person will do on a frontal lobe test.

Of course, polymorphic differences allow for only limited predictions about a person's overall performance. That's because many genes and many different alleles are responsible for most aspects of our behavior and psychological and cognitive functioning. So if someone announces the discovery of an "intelligence gene," don't believe it. Broader, more impressionistic profiles of some aspects of our personalities, however, will be possible. Richard Davidson, a psychologist at the University of Wisconsin-Madison, provides an example of what we can expect in regard to temperament.

Davidson has found that among infants as young as ten months, their response to forced separation from their mother correlates with activity in their prefrontal regions. Infants who cry or turn fussy show increased right-sided prefrontal activity compared to infants who don't get so upset. This is also true of adults. When shown films depicting positive or negative scenarios (a party versus a combat scene), adults with increased right side prefrontal activity are more likely to report distressing emotions than people with left-sided activity, who typically express a positive, upbeat approach to life.

Happy, extroverted people also show distinct activity in their amygdala when they encounter a happy face. Indeed, the degree of extroversion in a person correlates directly with the extent of amygdalar activity. When encountering a fearful or an angry face, in contrast, extroverts and introverts respond about the same. And this, of course, makes a good deal of sense. Self-preservation requires that all of us remain sensitive to other people's hostility and fear. But this tendency can be overdeveloped, as with abused children who respond to slight and often ambiguous expressions of anger. We know this because of the research of psychologist Seth Pollak, who is also at the University of Wisconsin-Madison.

At Pollak's Child Emotion Research Laboratory, children with a history of abuse were asked to look at a series of digitally morphed photos of facial expressions that ranged from happy to fearful, happy to sad, angry to fearful, or angry to sad. While some of the faces expressed a single emotion, most were blends of two emotions.

The responses of both abused and nonabused children didn't differ when reading expressions showing mostly happiness, sadness, or fear. But abused children tended to identify more faces as showing anger rather than fear or sadness. A photo with a blended expression of 60 percent fear and 40 percent anger tended to be interpreted by an abused child as showing only anger.

"It may be the case that physically abused children develop a broader category of anger because it's adaptive for them to notice when adults are angry," suggests Pollak.

Sensitivity to facial expressions of happiness, in contrast, isn't so directly linked to survival and therefore tends to vary according to personality traits, such as extroversion and introversion. Unfortunately this explanation isn't any help in deciding the neuroscientific version of the chicken or the egg question, Which comes first: the temperamental trait of extroversion, which is coupled, for unknown reasons, with an amygdala that readily responds to happiness in others? Or is the responsive amygdala the key here, with extroversion merely a consequence?

In the near future look for new and easily applied technologies to assess discrete cognitive abilities. For example, many perfectly normal people find it increasingly difficult to ignore distractions. In technical terms, they demonstrate what neuroscientists refer to as a deficit in "sensory gating." Think of your brain as a gatekeeper that continually evaluates the relevance of what's happening around you in light of your current goals and activities. As you read this sentence, your brain is operating as a gatekeeper to filter out background noise and distraction. But not everyone's gatekeeper functions efficiently.

Early sensory responses can also be used to predict the likelihood that a child's behavior will be similar to the parent's. For example, although not every child from an alcoholic family goes on to develop alcoholism, a sufficient number fall prey to the disorder that neuroscientists have long searched for a test to help identify those children who are at greatest genetic risk. Such a test is now available.

In the graph on page 203, notice the wave labeled P3. This positive, upward deflection occurs approximately 300 milliseconds after the volunteer hears a test sound. The amplitude, or height, of the P3 reflects the amount of information processed, while the latency (the time it takes for the P3 to occur) indicates processing speed. Increased P3 latency indicates a slowing of attention- or stimulus-evaluation time, whereas a decrease in P3 amplitude indicates decreased capacity for attention. Both amplitude and latency differ significantly in children who are at risk for alcoholism from children who aren't at risk. In fact, by looking at P3 responses a physician can distinguish between the adult children of alcoholics and those who come from families that are free of alcoholism. While variations in P3 latency and amplitude cannot predict that a *specific* individual will become an alcoholic, the test can serve as a vulnerability marker. Those with abnormal P3s should be encouraged to abstain from alcohol or, at the very least, moderate their drinking.

The P3 research underscores an important point that I have emphasized throughout this book: Over the next decade we can expect tests that provide important information about memory, language, and other aspects of human cognitive functioning. Moreover, these tests will help to provide a rational basis for selecting people best equipped to meet specific challenges. For example, people with an increased P3 latency and decreased amplitude should probably not be involved in occupations like air traffic control, which requires the rapid analysis of large amounts of data and quick decisions.

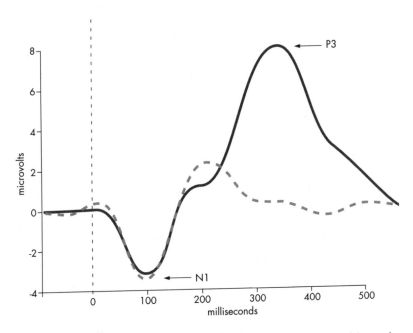

The height and length of the P3 wave in this diagram can be a predictor of those who are at risk for alcoholism.

Will Computers Be Able
to Model the Brain?

Such impressive advances in the neuroscience of cognition raise an important question: How many of the functions of the human brain will be equaled or bested by computers? According to Bill Gates, Microsoft Corporation chairman, none of us will live long enough to see the day when computers think and act like people. Gates believes that the brain is too complex to be compared to the logic gates and microcircuitry of a computer chip. "We don't understand the basic operations of the brain enough," he says.

Rather than trying to develop computers that exceed the powers of the brain, Gates suggests that we think of the computer as a tool that can be used in innovative applications. Thus, it's likely that computer advances over the next decade will be directed to augment the powers of the brain. In such collaborations, processing speed is likely to remain the computer's strongest suit. So far, even the most impressive computer performances against the human brain—such as the defeat of world chess champion Gary Kasparov by the IBM computer Deep Blue—rely on the computer's capacity to carry out millions of calculations per second.

As an example of what can be expected, Gates points to computers that can recognize almost any kind of handwriting, text, drawings, or annotations. Microsoft is also developing computers that can respond to gestures. And while such a computer could prove useful for people who are deaf or afflicted with other communication disabilities, the developers have already learned that programming a computer to distinguish a nod from a shrug is the easy part. More difficult is getting people to use the system, which according to Gates "is very tiring over the course of the day."

Implantation technologies provide another approach to enhancing the brain's power. While users may not find them tiring, they will find them more expensive and technologically advanced than the computer programs that attempt to simulate the brain. When implanted in the neck, the VNS Pulse Generator, for instance, stimulates the vagus nerve to forward electrical impulses to the brain. The device is already in use to prevent epileptic seizures and alleviate depression. Other presently available implanted devices lessen the tremors and movement impairments of Parkinson's disease.

Within the next five years radio-powered stimulators should be available to implant in individuals who will benefit from their ability to restore some movement to paralyzed limbs. And within ten years we can expect tiny microchips capable of mimicking the firing of neurons that have the potential to help repair damaged brains and spinal cords.

Most exciting are implants that will enable the blind and severely visually impaired to recover functional vision. I mentioned in chapter 8 the prosthetic sensory substitution devices that convert touch impulses into visual experiences. Other researchers, such as William Dobelle, developer of the Dobelle Institute's Artificial Vision System for the Blind, favor a more direct approach: implanting electrodes directly onto the visual areas of the brain. As with Paul Bach-y-Rita's system, it consists of a miniature camera mounted on eyeglasses and connected to a microcomputer. Instead of the information being converted to stimulate the skin, however, it proceeds directly to the brain. One patient, known as Jens, who lost vision in both eyes due to two separate accidents, recovered sufficient vision with the Dobelle artificial vision system to maneuver a Mustang convertible around the parking lot behind Dobelle's lab. "The next version may have enough resolution to use while driving in traffic," Dobelle told reporter Steven Kotler, who personally observed Jens's performance and reported on it for *Wired* magazine.

Other vision researchers, such as bioengineer Richard Normann of the University of Utah in Salt Lake City, are improving Dobelle's Artificial Visual System. Normann has reduced the size of the implant from the size of a thick quarter to about the size of a nail's head. And like a nail it is delicately "hammered" by a device into the cortical site where visual signals are processed. When perfected, the implant is expected to be sufficiently precise to stimulate individual neurons. It will also require less electrical current, decreasing the risk of overstimulating the brain and causing seizures. Less current also translates into greater visual resolution.

Further out on the horizon are implants capable of augmenting cognition itself. Theodore W. Berger, a professor of biomedical engineering at the University of Southern California in Los Angeles, is working on a computer chip that will combat Alzheimer's disease and other serious memory disorders by mimicking some of the functions of the hippocampus, the tiny seahorse-shaped cortical structure where memories are initially encoded. Berger has already created a mathematical model of the actions of neurons in the rat hippocampus and is now in the process of designing circuits to duplicate these actions. He plans on implanting the chips in the brains of rats or monkeys where, if everything goes according to plan, the chips will perform like a network of hippocampal neurons.

We Can Do It, but Should We?

All of these advances will bring more attention to the "value" questions inherent in brain research. Many of these topics we've explored throughout the book raise a host of ethical questions.

What are the implications for cherished concepts such as personal identity and human uniqueness when implanted bionic devices become able to restore or enhance the functions of the brain? Do we really want to further additional developments in technologies aimed at detecting when a person is lying? Should we support the development of tests to be given in infancy that will make it possible to predict adult intellectual capabilities? Should we modify our concepts of freedom and personal responsibility on the basis of findings that some criminals have abnormal brains? Questions such as these are providing the impetus for the development of a brand new area of neuroscientific exploration: neuroethics.

Neuroethics is concerned with the moral and ethical issues arising from new, brain-related scientific findings. It attempts to predict and respond to the social consequences of future advances in neuroscience. If a PET scan can reliably detect lying, should suspected criminals be forced to submit to the test? Should job applicants be compelled to take the test as a condition for employment? Obviously, such questions are not merely scientific but touch on many of the values that we hold dear in our democratic, free society.

Despite the undisputable value of subjecting neuroscientific advances to ethical scrutiny, it's probably naïve to think that neuroethics is going to provide easy answers to hard questions. Neuroethical dilemmas are certain to arise that will defy our most determined efforts at achieving satisfactory solutions to complex issues. In one example, consider the relationship between child abuse and later development of antisocial personalities and violence.

Experts have known for some time that children who suffer abuse are more likely than nonabused children to later become violent or exhibit criminal tendencies. Further, a child's age at the time of the abuse is somewhat predictive of future problems. Generally, the earlier in life children experience abuse, the more likely they will later develop behavioral problems. But whatever the child's age, abuse increases the overall risk of later criminality and/or violence by about 50 percent. Still, most maltreated children don't go on to become criminals. Why? And how does one distinguish those children who will later develop criminal or violent propensities?

It's likely that some abused children become violent or criminal because of genetic susceptibility factors. In essence, some potential future criminals may harbor a gene or genes that remain dormant unless "turned on" by abuse. One gene of particular interest codes for the enzyme monoamine oxidase A (MAOA), which breaks down several neurotransmitters in the brain including norepinephrine (NE), serotonin (5-HT), and dopamine (DA), rendering them inactive. Low MAOA levels in animals correlate with violence. Knocking out the MAOA gene in mice makes them more aggressive.

The first suggestion of a similar link between the MAOA gene and violence in humans came from a 1993 study of a Dutch family. Several men in this family had a defective MAOA gene (resulting in a deficiency of the enzyme) and were given to impulsive outbursts of violence. Little was made of this intriguing finding, however, since the family was small and the abnormal genetic mutation extremely rare.

In 2002 Terri Moffitt and Avshalom Caspi, both of King's College London and the psychology department at the University of Wisconsin-Madison, revisited the MAOA question. They turned to a study that began in 1972 that followed 1,037 children since birth. As anticipated, the researchers found that boys who had been abused were more likely to have become violent or criminal as adolescents and young adults. But researchers could target antisocial behavior with even greater precision by concentrating on abused boys with the genotype for low level of MAOA activity. Those with low MAOA activity were twice as likely to develop conduct disorders and three times as likely to commit a violent crime by age 26. In all, 55 males with histories of abuse and the low MAOA genotype made up only 12 percent of the study population but committed 44 percent of the crimes. In the absence of abuse, however, the low activity MAOA genotype didn't make the boys any more likely to be violent or antisocial.

With this as background, consider the following neuroethical dilemma: Imagine that you're a judge overseeing two trials of young men in their late twenties who have committed violent murders. You discover that both have histories of severe childhood abuse. You're aware that such abuse increases the odds of later violence or criminality, but since most abused children don't go on to become delinquents or criminals, you reason that the defendants must take full responsibility for their crimes.

At this point you learn that one of the defendants has the genotype indicating low MAOA activity, while the second defendant's MAOA activity is normal. Should the prisoner with the defective genotype be held less responsible for his crime than the man with a normal MAOA gene? In other words, should you consider genetic factors important enough to overshadow the traditional belief that people possess free will and therefore must be held accountable for their actions? If you do, then how would you proceed? Would you hold the prisoner with the genotype for low MAOA activity not guilty and set him free? Or would you consider him guilty but less so than his counterpart, and take the genetic information into account at the time of sentencing?

If you're not sure how you would resolve this dilemma, don't despair. I'm not sure how I would decide it either. And frankly, I'm not confident neuroscience is going to come up with answers to such ethical dilemmas. Despite whatever we may learn about the brain, a person's culpability for wrongdoing will always involve balancing multiple factors and deciding on the basis of incomplete information. Brain science isn't going to provide answers to questions that have eluded the best efforts of ethicists, criminologists, and philosophers.

Someone once compared the genes-environment relationship to that of a gun and its trigger. The genes are the loaded gun, and the environment is the agent that pulls the trigger. Genes, like guns, become operational only under the influence of the environment. It's a mistake, therefore, to focus exclusively on either genes or the environment.

"It's not at all just a simple result of a bad environment on the one hand or bad genes on the other," according to Antonio Damasio, head of neurology at the University of Iowa Carver College of Medicine. "It's a very complicated system that we're trying to understand. We're really in a quandary because we're entering uncharted waters."

Other uncharted waters involve "a whole series of physico-technologies that are going to be incorporated into our very flesh, that will become part of who we are," according to bioethicist Paul Wolpe. And as Richard Gregory and Paul Bach-y-Rita discovered with their blind subjects who regained vision, unpredictable emotional responses can occur in some people who were thought to benefit from attempts to help them. Some technological interventions may even bring about a fundamental revision in a person's self-concept.

Kevin Warwick, a professor of cybernetics at the University of Reading in England, had a chip implanted in his arm that enables coworkers in his laboratory to know his exact location at all times. Shortly after the chip implantation, Warwick began to "feel strangely connected to both the building and the computer. "I was more at one with the systems that were at work in my department," he said. He now hopes to convince his wife to have a chip implanted in her arm and eventually foresees chips placed in each of their brains in order to communicate telepathically. This is not simply wild fancy. The technology exists to make simple forms of telepathy an achievable goal. But what changes in the ways we think about ourselves will such a neuroscientific achievement bring about?

"We need to ask the broader question of what is the next step of evolution—using that word in its common sense rather than its technical sense—and how as a species we want to direct ourselves and towards what goals," according to Wolpe.

One thing is certain: Our brain's organization will undergo greater changes during the next several decades than at any time in our history. And technology will continue to be the compelling force behind those modifications. Most important, the changes in our brains brought about by technology will continue to provide us with the challenge of retaining our freedom and sense of identity while simultaneously utilizing soon-to-be-available techniques to vastly expand our mental horizons.

Bibliography

Bach-y-Rita, Paul. *Brain Mechanisms in Sensory Substitution*. New York and London: Academic Press, 1972.

Bach-y-Rita, Paul, Mitchell E. Tyler, and Kurt A. Kaczmarek. "Seeing with the Brain." *International Journal of Human-Computer Interaction*. In press.

Balter, Michael. "What Makes the Mind Dance and Count." *Science* 292 (June 1, 2001): 1636–1637.

Barber, Benjamin R. *Jihad vs. McWorld*. New York: Times Books, 1995.

Bates, John A., and Anil K. Malhotra. "Genetic Factors and Neurocognitive Traits." *CNS Spectrums: The International Journal of Neuropsychiatric Medicine* 7.4 (April 2002): 274–284.

Brown, W. E., S. Eliez, V. Menon, J. M. Rumsey, C. D. White, and A. L. Reiss. "Preliminary Evidence of Widespread Morphological Variations of the Brain in Dyslexia." *Neurology* 56 (March, 2001): 781–783.

Canli, Turhan, Heidi Sivers, Susan L. Whitfield, Ian H. Gotlib, and John D. E. Gabrieli. "Amygdala Response to Happy Faces as a Function of Extraversion." *Science* 296 (June 21, 2002): 2191.

Cardoso, Sylvia H. "Our Ancient Laughing Brain." *Cerebrum* 2.4 (Fall): 15–30. New York: Dana Press, 2000.

Cohen, R. A., R. F. Kaplan, D. J. Moser, M. A. Jenkins, and H. Wilkinson. "Impairments of Attention After Cingulotomy." *Neurology* 53 (September, 1999): 819–824.

Conrad, Peter. *Modern Times, Modern Places*. New York: Alfred A. Knopf, 1999.

Duncan, John. "An Adaptive Coding Model of Neural Function in Prefrontal Cortex." *Nature Reviews Neuroscience* 2 (November 2001): 820–829.

Ericsson, K. A. "Attaining Excellence through Deliberate Practice: Insights from the Study of Expert Performance." *The Pursuit of Excellence in Education*, edited by M. Ferrari, 21–55. Hillsdale, N.J.: Erlbaum, 2002.

———. "The Path to Expert Golf Performance: Insights from the Masters on How to Improve Performance by Deliberate Practice." *Optimising Performance in Golf*, edited by R. R. Thomas, 1–57. Brisbane, Australia: Australia Academic Press, 2001.

Gehring, William J., and Adrian R. Willoughby. "The Medial Frontal Cortex and the Rapid Processing of Monetary Gains and Losses." *Science* 295 (March 22, 2002): 2279–2282.

Gitlin, Todd. *Media Unlimited: How the Torrent of Images and Sounds Overwhelms Our Lives*. New York: Metropolitan Books/Henry Holt and Company, 2001.

Greene, Joshua D., R. Brian Sommerville, Leigh E. Nystrom, John M. Darley, and Jonathan D. Cohen. "An fMRI Investigation of Emotional Engagement in Moral Judgment." *Science* 293 (September 14, 2001): 2105–2109.

Helmuth, Laura. "Moral Reasoning Relies on Emotion." *Science* 293 (September 14, 2001): 1971–1972.

Johnson, Jeffrey G., Patricia Cohen, Elizabeth M. Smailes, Stephanie Kasen, and Judith S. Brook. "Television Viewing and Aggressive Behavior during Adolescence and Adulthood." *Science* 295 (March 29, 2002): 2468–2471.

Juergensmeyer, Mark. *Terror in the Mind of God: The Global Rise of Religious Violence*. Berkeley and Los Angeles: University of California Press, 2000.

Levitin, Daniel J. "In Search of the Musical Mind." *Cerebrum* 2.4 (Fall): 31–49. New York: Dana Press, 2000.

MacDonald, Angus W. III, Jonathan D. Cohen, V. Andrew Stenger, and Cameron S. Carter. "Dissociating the Role of the Dorsolateral Prefrontal and Anterior Cingulate Cortex in Cognitive Control." *Science* 288 (June 9, 2000): 1835–1838.

Miller, Greg. "The Good, the Bad, and the Anterior Cingulate." *Science* 295 (March 22, 2002): 2193–2194.

Rankin, Catharine H. "A Bite to Remember." *Science* 296 (May 31, 2002): 1624–1625.

Rayner, Keith, Barbara R. Foorman, Charles A. Perfetti, David Pesetsky, and Mark S. Seidenberg. "How Should Reading Be Taught?" *Scientific American* (March 2002): 85–91.

Shenk, David. *Data Smog*. New York: HarperCollins Publishers, 1998.

Shidara, Munetaka, and Barry J. Richmond. "Anterior Cingulate: Single Neuronal Signals Related to Degree of Reward Expectancy." *Science* 296 (May 31, 2002): 1709–1711.

Simos, P. G., J. M. Fletcher, E. Bergman, J. I. Breier, B. R. Foorman, E. M. Castillo, R. N. Davis, M. Fitzgerald, and A. C. Papanicolaou. "Dyslexia-Specific Brain Activation Profile Becomes Normal Following Successful Remedial Training." *Neurology* 58 (April, 2002): 1203–1213.

"What Are the Costs of Multitasking?" *Neurology Reviews* 9.9 (September 2001): 59–60.

Index

Boldface page references indicate illustrations.

Columbia University (New York City), 143
Computers, modeling of brain by, 204–7
COMT (catechol O-methyltransferase), 199
Conditioning, fear, 66
Conflict, 115–17
Constraint Induced (CI) Movement Therapy, 150–53
Corpus callosum, 1, 64, 99
Cortical reorganization, 149–50
Crawlers, across television screen, 39–40
CREB, 143
Cross-modal matching, 175

D

Daydreams, 81
Decision making
emotional component of, 112
impulsivity, 119
Depression
hippocampal atrophy and, 122–23
medicalization of, 135–36
in terrorist attack aftermath, 74
treatment by
psychopharmacology, 121–24
rTMS, 189
talk, 121
VNS Pulse Generator, 205
Desensitization, to pain and suffering of others, 85
Detachment, feelings of, 79
Distraction
ADD/ADHD, 42–46
blocking with music, 61
by crawlers across television screen, 39–40
efficiency reduction by, 57

feelings of, 41
of multitasking, 45
response to, 201
DNA chip, 127–30, 144
Dobelle Institute, 205–6
Dopamine-boosting drugs, for memory enhancement, 142
Driving
cell phone use during, 56, 59
emotional contagion and, 37
performance, 18–19
Drowsiness, counteracting, 133–34
Drugs
addiction and anterior cingulate cortex, 117
lifestyle, 132–34
personalized medications, 127–32
Dynamic Brain Atlas, 190–91
Dynamic vigilance, 115
Dyslexia, 184–86
Dysthymic disorder, 137
Dystonia, 156–58

E

EEGs, of musicians, 101
Efficiency, reduction by multitasking, 54–59
Electrical activity, of the brain
Brain Fingerprinting and, 106–8
event-related brain potentials (ERPs), 109
lie-detection and, 118–19
speed of, 110–11
Emotional contagion, 37
Emotional shutdown, 79–80
Emotions
decision making, effect on, 112
disturbing images and, 69–77
imaging research on, 111–12, 114
influence on reasoning, 112, 113–15